Mayuri Napagoda and Lalith Jayasinghe (Eds.)
Chemistry of Natural Products

Also of interest

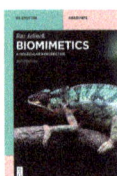

Biomimetics.
A Molecular Perspective
Jelinek, 2021
ISBN 978-3-11-070944-5, e-ISBN (PDF) 978-3-11-070949-0

Pharmaceutical Chemistry.
Vol 1: Drug Design and Action
Campos Rosa, Camacho Quesada, 2017
ISBN 978-3-11-052836-7, e-ISBN (PDF) 978-3-11-052848-0

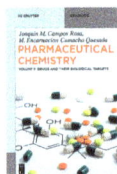

Pharmaceutical Chemistry.
Vol 2: Drugs and Their Biological Targets
Campos Rosa, Camacho Quesada, 2017
ISBN 978-3-11-052851-0, e-ISBN (PDF) 978-3-11-052852-7

Chemical Drug Design
Kumar Gupta, Kumar (Eds), 2019
ISBN 978-3-11-037449-0, e-ISBN (PDF) 978-3-11-036882-6

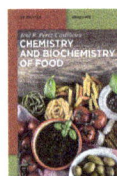

Chemistry and Biochemistry of Food
Jose R. Pérez-Castiñeira, 2020
ISBN 978-3-11-059547-5, e-ISBN (PDF) 978-3-11-059548-2

Chemistry of Natural Products

Phytochemistry and Pharmacognosy of Medicinal Plants

Edited by
Mayuri Napagoda and Lalith Jayasinghe

DE GRUYTER

Editors
Prof. Mayuri Napagoda
Department of Biochemistry
Faculty of Medicine
University of Ruhuna
Labuduwa Rd.
Galle
Sri Lanka
mayurinapagoda@yahoo.com

Prof. Lalith Jayasinghe
Natural Products Research Division
National Institute of Fundamental Studies
Hantana Rd.
Kandy 20000
Sri Lanka
ulbj2003@yahoo.com

ISBN 978-3-11-059589-5
e-ISBN (PDF) 978-3-11-059594-9
e-ISBN (EPUB) 978-3-11-059365-5

Library of Congress Control Number: 2022930255

Bibliographic information published by the Deutsche Nationalbibliothek
The Deutsche Nationalbibliothek lists this publication in the Deutsche Nationalbibliografie;
detailed bibliographic data are available on the Internet at http://dnb.dnb.de.

© 2022 Walter de Gruyter GmbH, Berlin/Boston
Cover image: altmodern/iStock/Getty Images Plus
Typesetting: Integra Software Services Pvt. Ltd.
Printing and binding: CPI books GmbH, Leck

www.degruyter.com

About the editors

Prof. Mayuri Napagoda completed her B.Sc Special Degree (Botany) with a First Class honours at the University of Colombo, Sri Lanka in 2001 and Graduateship in Chemistry at the Institute of Chemistry Ceylon in 2003. She obtained her M.Phil (Natural Products Chemistry) from University of Peradeniya, Sri Lanka (2005) and PhD (Natural Products Chemistry) from Friedrich-Schiller-University, Jena, Germany (2014). She received a number awards including the International Max Planck Research School (IMPRS) fellowship (2010-2013); DAAD research fellowship (2019-2020); Hiran Tillekeratne Award for the Outstanding Postgraduate Research in Medicine -University Grants Commission, Sri Lanka (2008); Kandiah Memorial Award from Institute of Chemistry Ceylon (2006) for the best piece of research by a postgraduate student; Swarna Senathirajah Memorial Award for Academic Excellence-University of Colombo, Sri Lanka (2001) and Institute of Chemistry Merit Award (1998). She is an expert in the field of natural products chemistry and secured many national and international research grants, travel grants, and research excellence awards. She serves as a manuscript reviewer of several journals and a reviewer of research grant applications. Her research interests are not limited to ethnobotanical studies and bioactivity studies of medicinal plants, but extended towards the application of nanotechnology in herbal medicine and herbal cosmetics. She is presently serving as a Professor at the Department of Biochemistry, Faculty of Medicine, University of Ruhuna, Sri Lanka.

Prof. Lalith Jayasinghe obtained his BSc (Hons) in Chemistry from University of Peradeniya, Sri Lanka in 1985; Ph.D. (Natural Product Chemistry) from University of Peradeniya, Sri Lanka in 1992 and Diploma in Natural Products Chemistry, TIT Japan in 1994. He is an Alexander von Humboldt Research Fellow at University of Hohenheim (1999/2000) & Jacobs University Bremen (2011, 2015 & 2019) Germany. He has been awarded following fellowships and awards; IPICS Fellowship in 1988 to University of Karachi; Kandiah Memorial Award-Institute of Chemistry Ceylon (1992); TWAS-NARESA Young Scientist Award-1992; L. De Silva Gold Medal (2021), Institute of Chemistry Ceylon and UNESCO-Mombusho Research Fellowship (1993). He is a visiting Scientist of TIT, Japan (2004); University of Mississippi, USA (2004 - 2005); University of Milan, Italy (2006) and a Visiting Professor of TIT, Japan (2009) and Meiji University, Japan (2019). He is a Fellow of the National Academy of Science of Sri Lanka (2012). He received Presidential Award for Scientific Publication for 8 Years. He is presently serving as a Senior Research Professor, National Institute of Fundamental Studies, Kandy, Sri Lanka.

https://doi.org/10.1515/9783110595949-202

Preface

Plants produce secondary metabolites for a plethora of roles and humans have always inquisitively attempted to harness these benefits. Thus, applications of natural products chemistry have become all-pervasive in modern society and especially in the fields of medicine and pharmacology. About half of the drugs currently in clinical use are based on natural product scaffolds. Therefore, a deeper understanding of secondary metabolites present in medicinal plants and their biosynthesis, biological activities as well as isolation and separation techniques is essential for researchers in this field. This book provides an easy-to-read overview of secondary metabolites in medicinal plants, their pharmacological potential, and current techniques involved in the isolation and structure elucidation of bioactive natural products. In addition, the readers get the opportunity to learn about poisonous plants and some important aspects relevant to the safety and quality of herbal medicinal products. Hence, this book will be useful for undergraduate and postgraduate students as well as other researchers in the field of natural products chemistry and pharmaceutical industry.

https://doi.org/10.1515/9783110595949-203

Contents

List of contributors

Chandani Ranasinghe
Department of Chemistry
The Open University of Sri Lanka
Nawala, Nugegoda, Sri Lanka
crana@ou.ac.lk
Chapter 3

Mayuri Napagoda
Department of Biochemistry
Faculty of Medicine, University of Ruhuna
80000 Galle
Sri Lanka
mayurinapagoda@yahoo.com
Chapters 1, 8, 9, and 10

D. S. A. Wijesundara
National Institute of Fundamental Studies
20000 Kandy
Sri Lanka
Chapter 1

Thushara Diyabalanage
Actives International LLC
6 Pearl Court, Unit G
Allendale, NJ 07401
USA
thusharad@activesinternational.com
Chapter 2

K. G. N. P. Piyasena
Tea Research Institute of Sri Lanka
Talawakelle, Sri Lanka
nelumpriya@yahoo.com
Chapter 6

M. M. Qader
Department of Chemistry
The Open University of Sri Lanka, Nawala
Nugegoda, Sri Lanka
Chapter 6

Sathya Sambavathas
Faculty of Allied Health Sciences
University of Jaffna, Jaffna, Sri Lanka
Chapter 7

Nilupa R Amarasinghe
Faculty of Allied Health Sciences
University of Peradeniya, Peradeniya
Sri Lanka
Chapter 7

Lalith Jayasinghe
National Institute of Fundamental Studies
Hantana Road, Kandy, Sri Lanka
ulbj2003@yahoo.com
Chapters 7 and 8

Yoshinori Fujimoto
National Institute of Fundamental Studies
Hantana Road, Kandy, Sri Lanka
And
School of Agriculture, Meiji University
Kawasaki, Kanagawa 214-8571, Japan
Chapter 7

Jagadeshwar Reddy Thota
CSIR-Indian Institute of Chemical Technology
Tarnaka, Hyderabad 500 007, Telangana
India
redditj@gmail.com
Chapter 11

Ravi Kumar Maddula
CSIR-Indian Institute of Chemical Technology
Tarnaka, Hyderabad 500 007, Telangana
India
Chapter 11

Sukanya Pandeti
CSIR-Indian Institute of Chemical Technology
Tarnaka, Hyderabad 500 007, Telangana
India
Chapter 11

Naga Veera Yerra
CSIR-Indian Institute of Chemical Technology
Tarnaka, Hyderabad 500 007, Telangana
India
Chapter 11

https://doi.org/10.1515/9783110595949-205

Kanchana Wijesekera
Faculty of Allied Health Sciences
University of Ruhuna, Galle, Sri Lanka
kdwijesekera@gmail.com
Chapters 4 and 5

Aruna S. Dissanayake
Faculty of Allied Health Sciences
University of Ruhuna, Galle, Sri Lanka
Chapter 4

Sewwandi Subasinghe
Faculty of Allied Health Sciences
University of Ruhuna, Galle, Sri Lanka
Chapter 5

Chamika Liyanaarachchi
Faculty of Medicine
University of Ruhuna, Galle 80 000
Sri Lanka
Chapter 8

Sanjeeva Witharana
Faculty of Engineering
University of Moratuwa, Moratuwa 10400
Sri Lanka
Chapter 8

Part I: **General**

Mayuri Napagoda, D. S. A. Wijesundara

1 Medicinal plants as sources of novel therapeutics: the history, present, and future

1.1 Introduction

Plants form the basis of various traditional and folk medicines that have been in practice for thousands of years. Even today, plants are considered as a rich source of therapeutic agents for the treatment and prevention of diseases. According to the World Health Organization (WHO), approximately 80% of the inhabitants in the world depend mainly on traditional medicines for their primary health care. It is estimated that at present, more than 35,000 plant species are employed for medicinal purposes [1].

A medicinal plant is usually described as "any plant which, in one or more of its organs, contains substances that can be used for therapeutic purposes or which are precursors for the synthesis of useful drugs" [2]. These plants also become the source of natural products that can be developed into novel drugs or can be utilized as drug leads. Noteworthy, up to 50% of the approved drugs during the last 30 years are from either directly or indirectly from natural products. For example, in the area of cancer, out of the approved 175 small molecules over the time frame from the 1940s to 2014, 85 (49%) were either natural products or natural product derivatives [3]. Besides, natural dietary supplements are also gaining much popularity among the general public. Particularly, cancer patients in the USA have started to use new dietary supplements with natural ingredients after being diagnosed with cancer. These herbal medicinal products are available as single isolated/enriched compounds or as complex mixtures of several biologically active compounds. Further, these could be obtained from a single herb or combination of herbs, as polyherbal formulations and are prepared in different ways like decoction, tinctures, teas, syrups, essential oils, ointments, salves, and tablets/capsules with the powdered form of the whole plant/plant part or dried extract [4].

The increasing interest in medicinal plant research is clearly reflected by the number of recent publications that have increased more than threefold from 2008 (4,686 publications) to 2018 (14,884 publications). Fitzgerald et al. [5] revealed that the largest proportion of publications cited in current databases over the last 10 years are in the disciplines of pharmacology and pharmacy and it is followed by plant sciences, biochemical molecular biology, and agriculture research. Moreover,

Mayuri Napagoda, Department of Biochemistry, Faculty of Medicine, University of Ruhuna, 80000 Galle, Sri Lanka, e-mail: mayurinapagoda@yahoo.com

D. S. A. Wijesundara, National Institute of Fundamental Studies, 20000 Kandy, Sri Lanka

https://doi.org/10.1515/9783110595949-001

the majority of those publications have emerged from China, India, the USA, and South Korea, indicating the strong medicinal plant traditions in Asia as well as the USA's dominant presence as an international user of herbal products [5].

Thus this chapter gives a brief overview of the history of medicinal plants, the challenges faced in the development of herbal-based drugs, and some future prospects in the field of herbal medicine.

1.2 The use of medicinal plants: historical perspective

The relationship between plants and humans has been existing since time immemorial. Early humans exploited the plants around them for use as food, fuel, clothing, shelter, and medicine [6]. Fossil records indicate that prehistoric humans had used plants as medicine at least 60,000 years ago in the Middle Paleolithic age [7, 8]. It is assumed that the treatment of open wounds had included the cleaning and packing with plant parts or plant extracts, some of which might be beneficial in cleansing and healing wounds. Among the objects found with the mummified body of Ötzi the Iceman who lived about 5,300 years ago, there were woody fruits of a bracket fungus *Piptoporus betulinus*. It has been revealed that *P. betulinus* contains toxic resins and agaric acid which are powerful purgatives, along with oils that are toxic to metazoans and with antibiotic properties [9].

It is speculated that the pharmacological knowledge of primitive man might have come from experimentation and sometimes they might have judged the use and purpose of plants just by examining what the plant resembled. For example, the black speck in the flower of the plant eyebright (*Euphrasia officinalis*) appears as a pupil in the eye and thus was used for diseases in the eye. Similarly, plants with bright yellow flowers were used against jaundice in which the white parts of the eye get turned into yellowish color [10].

The oldest written evidence of the use of medicinal plants for the preparation of drugs has been found on a Sumerian cuneiform tablet which is believed to be from 3000 BC. Fifteen pharmaceutical prescriptions composed of milk, snakeskin, turtle shell, *Cassia*, myrtle, asafetida, thyme, willow, pear, fig, fir, and date have been described there, although it lacks the context on associated diseases or the amounts of the ingredients. Interestingly, all parts of plants had been used for those prescriptions. Narcotics derived from *Cannabis sativa* (hemp), *Mandragora* spp. (mandrake), *Lolium temulentum* (darnel), and *Papaver somniferum* (opium) were utilized by the ancient Mesopotamians [10, 11].

Traditional Chinese Medicine dates back to about 2500 BC and the oldest medical writings on herbs described dozens of herbs in a variety of situations related to healing and diet [12]. The Chinese book *Pen T'Sao*, written by Emperor Shen Nung around 2500 BC, on roots and grasses revealed 365 drugs composed of dried parts of medicinal plants like yellow gentian, ginseng, cinnamon bark, camphor, *Podophyllum*, jimson weed, and *Ephedra* [13, 14].

The traditional Indian medicine, or Ayurveda, developed significantly during the Vedic period (2500–600 BC) and the descriptions of the system are available in ancient literature such as Rig-Veda, Yajur-Veda, and Atharva-Veda, which mention the utilization of plants for treatment purposes. Three groups of plants have been recognized in Rig-Veda as trees (Vriksha), herbs (Osadhi), and creepers (Virudh) while the shape and morphology of plants were also described in Atharva-Veda. Four groups of medicinal plants were described in Yajur-Veda [13, 15–17]. The Caraka-Samhita (Compendium of Maharishi Caraka) and Sushruta Samhita, dating to the period of 900–600 BC, are two fundamental texts on Indian traditional medicine and describe hundreds of pharmacologically active herbs and spices [17].

The Egyptian pharmaceutical record "Ebers Papyrus," written circa 1550 BC, is the most complete and most famous medical papyri. It describes hundreds of magical formulas and folk remedies refereeing to about 700 plant species including pomegranate, castor oil plant, garlic, onion, *Aloe*, *Senna*, fig, willow, coriander, juniper, etc. [13]. Plant extracts were prepared and either taken internally or applied topically, while some were administered by fumigation and vapor inhalation [18].

The Egyptian tradition was transmitted to Greek and Roman medical systems over time triggering the use of plant species against various ailments. Greek philosopher Aristotle (384–322 BC) has described 500 crude drugs employed in the treatment of different pathological conditions, while the Greek Physician Hippocrates (460–370 BC), the father of modern medicine, believed that disease had natural causes; thus, various herbal remedies were used in his treatments. He mentioned about 400 medicinal substances of herbal origin [15, 19].

Theophrastus (370–287 BC), who is considered as "the father of Botany," wrote two books *"De Causis Plantarium"* – Plant Etiology and *"De Historia Plantarium"* – Plant History. In these books, he included a classification of more than 500 medicinal plants known at the time [15]. Further, Theophrastus described on the season and the method for the gathering of useful medicinals; for example, the best juices are collected in the summer, while spring or autumn would be the best time to gather the most useful roots [20].

The Roman writer Cornelius Celsus (25 BC–50 AD), who wrote the book *De Medicina*, described the preparation of numerous ancient medicinal remedies and quoted about 250 medicinal plants such as *Aloe*, poppy, pepper, cinnamon, the star gentian, and cardamom [13]. In around 60 AD, the Greek physician Pedanius Dioscorides (40–90 AD) documented over 600 curative plants in his book *De materia medica* which formed the core of the European pharmacopeia. Chamomile, garlic, onion, ivy, nettle, sage, coriander, parsley, willow, etc. are some of the most appreciated domestic plants described by Dioscorides. The descriptions of the medicinal plants included their outward appearance, locality, mode of collection, preparatory methods, and the therapeutic effects [13]. Similarly, Pliny the Elder (23–79 AD) introduced *Naturalis Historia*, a work that includes myths and folklore, trees, and medicinal plants [21].

Claudius Galen (129–199 AD) introduced the concept of pharmaceutical formulation to formulate stable and therapeutically effective drugs and published at least 30 books on plants [15, 22]. He also introduced several new plant drugs that had not been described by Dioscorides, for example, Uvae ursi folium, which had been used as an uroantiseptic and a mild diuretic [13].

During the Middle Ages, the monasteries preserved medical knowledge in Europe where monks who were in their monasteries planted and experimented on the species described in classic texts. Meanwhile, the Arabic scholars translated many classical Greek texts into Arabic and complemented it with their own medicinal expertise, as well as the knowledge of herbs from Chinese and Indian traditional medicines [23]. The Persian pharmacist and the physician Avicenna wrote "Canon Medicinae" and "Kitab Ash-Shifa," while Ibn al-Baitar recorded hundreds of medicinal plants in his "Corpus of Simples" [15, 21, 22]. Moreover, the toxic aspects of various plants were also described by Arabs, for example, *The Book on Poisons and Antidotes* by Abu Musa Jabir ben Hayyan [24].

"The Black Death," which is considered as one of the most devastating pandemics in human history, had swept through Europe in the thirteenth and fourteenth centuries. As the physicians were not knowledgeable at that time to deal with the infection, superstitious practices like burning aromatic herbs and bathing in rosewater or vinegar were also performed [21].

Although the emphasis paid on herbal sciences had declined during the late Middle Ages, several herbalists fostered this field, especially during the sixteenth century. In the dawn of Renaissance, Paracelsus (1493–1541 AD) reintroduced opium for medical use in Western Europe [25]. There was no concept of the geographical distribution of plants in the early sixteenth century, and Leonhart Fuchs (1501–1566 AD) became the first herbalist to describe the American introduction of previously unknown plants into Europe. His book *De historia stirpium* covers 497 native European and introduced plants and over 500 woodcut illustrations [26]. His work became a masterpiece and considered as the standard scholarly study on plants until Carolus Linnaeus (1707–1788 AD) introduced the new taxonomy, the binomial system [27]. Meanwhile, in England, John Gerard (1545–1612 AD) published *Herball* or *Generall Historie of Plantes*, and in 1618, London Pharmacopoeia was compiled using previous work on the medicinal plants [22].

Until the nineteenth century, medicinal plants were employed on an empirical basis, neither with mechanistic knowledge on their pharmacological activities nor with the active constituents [23]. The early nineteenth century was a turning point in the field of herbal medicine as attempts were made for the isolation of the active principles of commonly used plants such as poppy, belladonna, autumn crocus, and Saint-Ignatius' bean. These isolations were then followed by the commercialization of morphine, the first commercial pure natural product in 1826; the aspirin, the first semi-synthetic pure drug based on a natural product in 1899; and many other pharmaceutically important natural products thereafter [22].

Despite the advent of other drug discovery approaches like molecular modeling and combinatorial chemistry, the impact of natural products as new clinical candidates in the drug discovery programs is still very high. For example, 1,073 new chemical entities belonging to the group of small molecules had been approved between the period of 1981 and 2010 and more than half of those were based on natural product scaffolds. Interestingly, a substantial number of these compounds were from higher plants [23]. Thus natural-product-derived compounds are still proving to be an invaluable source of medicines for humans and the indigenous knowledge on medicinal plants play a vital role in expanding the horizons of the modern pharmaceutical industry. In this respect, ethnobotanical studies could be indispensable tools for gathering folklore knowledge.

1.3 Ethnobotany in drug discovery: pros and cons

The term "ethnobotany" was first introduced in 1896 as "the study of plant use by humans" by an American botanist John Harshberger. Thus it studies various aspects of how plants are used by people as food, cosmetics, textiles, and medicines including all the beliefs and cultural practices associated with their use [28]. On the other hand, the more recently introduced term "ethnopharmacology" describes a multidisciplinary area of research, concerned with the observation, description, and experimental investigation of indigenous drugs and their biological activities [29]. Ethnobotany has undergone a radical transformation during the last two decades [30].

Leopold Glück, a German physician working in Sarajevo, published his work on traditional medical uses of plants among the rural people in Bosnia in 1896, and it is believed that this work would be the first modern ethnobotanical work [31]. Since then a large number of ethnobotanical studies were carried out in different regions around the globe and some of the notable work includes those of Richard Evans Schultes and his students such as Wade Davis and Mark Plotkin in the South American Amazon [32].

Ethnobotanical studies play an important role in the preservation of traditional knowledge through proper documentation. A study of a rural population in Argentina revealed that for the transmission of knowledge of medicinal and edible plants, family members (especially mothers) play a major role while experienced traditional healers outside the family also made a great contribution [33, 34]. As smaller and more vulnerable tribes and indigenous groups become increasingly fragmented and threatened by modern development pressures in developing countries, it is feared that folk knowledge might get lost forever [34]. Also in some communities, the wealth of knowledge is rapidly diminishing not only due to the dearth of elderly people who are knowledgeable on traditional healing systems but also due to the lack of interest in the younger generation to acquire this knowledge systematically [35]. Also, dramatic destruction of ecosystems and the ruthless use and overexploitation of medicinal plants solely for commercial purposes compel to accelerate studies of ethnomedicine along with biomedical and phytochemical studies for the development of new natural products and drugs needed by humans [34, 35].

Ethnobotanical studies are proven to be an effective approach to reveal the hidden potential of plants against various illnesses and thereby could contribute toward the drug discovery programs by providing information on the selection of plants or specific phytochemicals to be tested in experimental models of various diseases. On several occasions, the results of ethno-directed investigations have been compared with random search for plants for specific therapeutic purposes. According to Khafagi [36], 83% of the plants in Sinai, Egypt selected using an ethnobotanical approach elicited antimicrobial activities while only 42% of the randomly selected plants exhibited the bioactivity [36]. Similarly, Slish et al. [37] reported that 4 out of 31 plants selected in Belize using the ethno-directed study displayed vascular smooth muscle relaxant activity; however, none of the randomly collected 32 plants exhibited this property [37].

However, there are number of pitfalls associated with ethnobotanical/ethnopharmacological studies, particularly concerning the design of studies and collection and interpretation of data. This demands proper training and sound knowledge of the international literature from investigators of theoretical and methodological contributions to this field. Besides, the selection of plants relevant for bioprospecting based on their popularity and usefulness is sometimes doubtful, while the exclusion of information essential for efficiently testing the plants is another error observed in the latest ethnobotanical studies. In order to overcome these limitations, it is recommended that researchers should clearly establish the goals of their study, for example, whether they are going to study one single and well-defined therapeutic activity, or the full range of knowledge of the local medical system. Further, during the selection of informants for the study, the age, gender, and social function of the individuals should be considered, particularly the role of women and elders who are supposed to possess greater knowledge on medicinal plants due to their role in the home and family care, and their longer interaction with the environment respectively. Also, the researchers should keep in mind that the high presence and importance of a particular plant in a local healing system might not always be linked to its pharmacological effect, whereas plants that are mentioned less frequently might be important for bioprospecting; thus, low popularity does not necessarily mean lack of efficacy. Because of the cultural validation and the local belief in its effectiveness, a widely popular plant may act like a "placebo," despite the absence of biologically active compounds. Moreover, the plant species located geographically closer to a local community may be used more often, thus imparting greater importance. On the other hand, a plant that has been rarely mentioned could be a recent introduction to a local medical system, or else, the knowledge on its healing potential might have been restricted to a few families or individuals as a family secret or has a low availability in the study area concerned. Therefore, such rarely mentioned plants might actually be highly valuable from the bioprospecting perspectives [38].

Although the "ethnobotanical approach to bioprospecting" has resulted in the development of at least 88 new pharmaceuticals like the muscle relaxant tubocurarine and the antimalarial drug quinine, there were instances where this approach was not as effective as it was anticipated. A well-known example was the project conducted by Shaman Pharmaceuticals in South San Francisco, California, USA, with the vision "collaborating with the rainforest's indigenous people as part of a sophisticated drug discovery and development process of modern Western medicine" [39]. A team of botanists and physicians were sent to 30 countries to work directly with indigenous communities and to interview the traditional healers to learn about the plants that are used to treat illnesses and how the patients are treated. Although the initial interest was directed toward antifungal and antiviral agents, the active compounds discovered were failed in the clinical trials; thus, the efforts were made to assess the antidiarrheal activity. "SP-303," a mixture of proanthocyanidin polymers isolated from the latex of *Croton lechleri* was found to be clinically efficacious and developed as a dietary supplement for diarrhea. It has been realized that the applications were different from indigenous ethnobotanical uses of *Croton* sap. Despite the collection of 1,000 plants and screening of 800, of which 420 identified with biological activity and leading to 20 patents, the Shaman project is considered as a victim of bad timing in its choice of the search strategy. The failure of the project signifies the need for new models of these approaches to drug development as well [7, 39].

The researchers involved in the collection of ethnobotanical knowledge of indigenous people should be aware of Convention on Biological Diversity (CBD) and intellectual property rights; thus, it is recommended to obtain prior informed consent for the use of the resources of those indigenous people and their traditional knowledge. Gaining the consent of indigenous people is a time-consuming process that involves the identification of appropriate indigenous communities to work with and getting their approval for sharing knowledge and resources as well as negotiation of appropriate contracts and compensation packages [39, 40].

Because of the complicated issues associated with ethnobotanical research, pharmaceutical companies prefer to use literature and database search rather than conducting ethnobotanically directed search itself [39]. NAPRALERT is one of the famous databases that were designed to evaluate the literature on natural products for the identification of new sources of commercially significant or clinically useful drugs. This database contains data on upward of 60 000 species, including more than 200,000 distinct chemical compounds of natural origin and 90,000 reports of ethnomedical uses of plants as well as other organisms. More than 770,000 unique pharmacological records are there representing more than 4,000 different pharmacological activities. These data have been extracted from over 200,000 scientific articles and reviews from approximately 10,000 scientific journals. Thereby it provides essential information to researchers who are engaged in medicinal plant research and drug development as well as the botanical dietary supplement industry [41].

There are several other herbal medicine databases with scientific data on the use and study of herbs for health, namely, the herb information knowledgebase (THINKherb) database, Traditional Chinese medicine information database (TCM-ID), Traditional Chinese medicine integrated database (TCMID), and Indian Plant Anticancer Compounds Database (InPACdb). THINKherb contains 499 herbs, 1,238 genes involving human, mouse, and rat, 825 diseases, 245 pharmacological activity, and 373 signaling pathways. TCM-ID composed of 1,588 prescriptions, 1,313 herbs as well as 5,669 herbal ingredients along with the 3D structure of 3,725 herbal ingredients. TCMID contains 47,000 prescriptions, 8,159 herbs, 25,210 compounds, 6,828 drugs, 3,791 diseases, and 17,521 related targets. On the other hand, InPACdb provides comprehensive information on anticancer activity of the phytochemicals of Indian origin. As of recent times, these databases turned out to be a valuable resource for drug development and drawn the attention of researchers in both academia and industry [1].

1.4 From plants to the pharmacy shelf: the drug development process

The development of new drugs from the plant sources is a complex, time-consuming, and expensive process (Figure 1.1) as it is carried out in three elaborate steps, namely, pre-drug stage, quasi drug stage, and full drug stage. The first stage of drug development is the pre-drug stage and involves the information-driven selection of plants either based on indigenous use or from the results obtained in animal studies. Then in the quasi drug stage, the extracts are prepared, phytochemicals are screened, and the structure and composition are elucidated. Further, the bioactivity evaluations are conducted for the identification of possible lead compounds. If necessary, the lead compounds are subjected to structural modifications as well. Once the lead compound is identified, it is structurally modified if needed. Thereafter, it is evaluated in animal models, *in vitro* studies, and clinical trials, and upon the approval, it enters as a marketed drug [4].

The new drug candidate (irrespective of whether a phytochemical or not) must undergo pre-clinical trials followed by different stages of clinical trials, i.e., Phase 0 (optional), Phase I, Phase II, Phase III, and Phase IV. Preclinical studies are required before the initiation of the clinical trial. These pre-clinical studies involve *in vitro* and animal experiments (*in vivo*) at different doses of the study drug to determine the pharmacodynamics, pharmacokinetics, and toxicology of the drug.

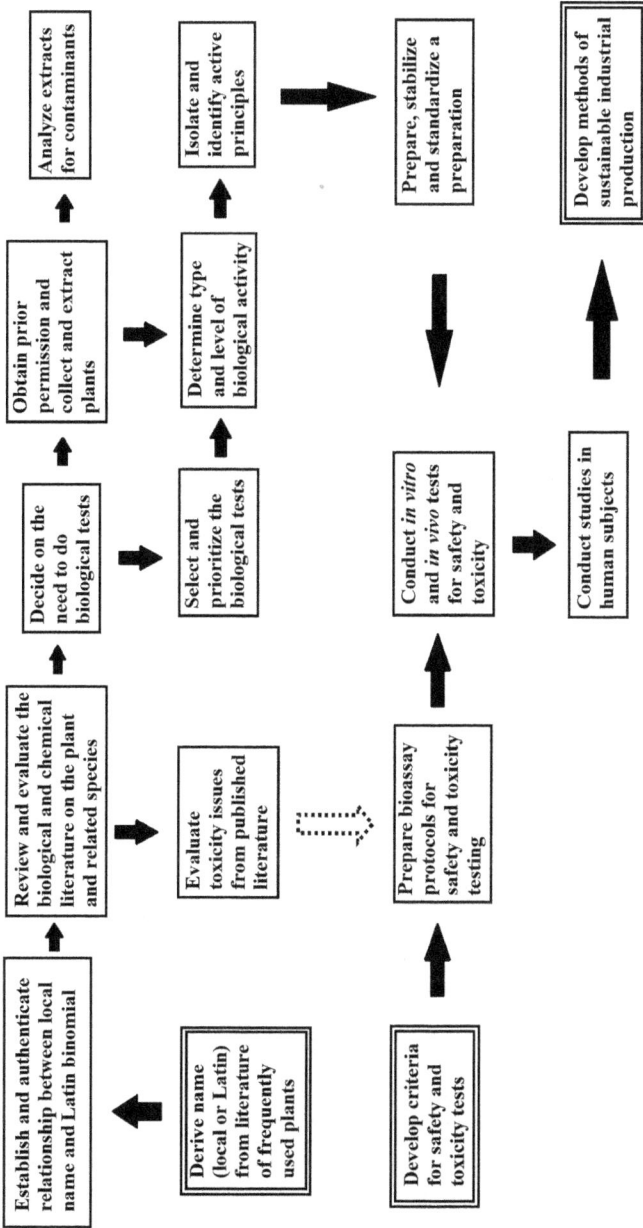

Figure 1.1: A flow chart for the study of plants used in traditional medicine (adapted from Cordell and Colvard, 2005 [40]).

Phase 0 experiments are optional exploratory trials where a small group of individuals, usually 10–15, are tested with a single, sub-therapeutic dose to gather pharmacokinetic information, thus to access whether the drug candidate performs as expected to take forward into further development. Then in Phase I studies, the new drug is administered to 20–100 normal healthy individuals to determine the maximum tolerated dose of the drug, the common and serious adverse effects of the drug, and also its pharmacological, pharmacodynamic, and pharmacokinetic properties. Nearly 70% of drug candidates move from Phase I to Phase II. Phase II studies are conducted with several hundred patients with specific diseases. The study population is well defined by inclusion and exclusion criteria, and based on the dose or dose range determined in Phase I, dose-response in patients and the drug's biological activity and efficacy are evaluated. Approximately one-third of tested new drugs move into Phase III trials which are carried out using a large number of patients and with specific diseases to determine efficacy, effectiveness, and long-term side effects. Usually, about 25% to 30% of the new drug candidates progress to the next phase, Phase IV. Phase IV trials are long-term studies involving more than 10,000 individuals of the relevant patient population and normally conducted after the approval of the regulatory agency. This study aims to assess the drug's real-world effectiveness. In some instances, the outcome of Phase IV studies could lead to a withdrawal of the drug from the market or for a restriction to particular uses [42].

Despite the fact that many botanicals are in the pipeline of clinical trials, only a few have ended up as a commercial drug, hence the rest failed at different stages of the clinical trial. One example is a different formulation of "SP-303" identified in the project undertaken by Shaman Pharmaceuticals. This formulation called "Virend" was developed as a topical formulation for the treatment of genital herpes, and it progressed to the clinical trials from the pre-clinical stage within a short period, i.e., after 24 months of laboratory testing. However, Shaman Pharmaceuticals halted the further development of Virend when it demonstrated no additional benefit over the existing drug for herpes, acyclovir [39]. Similarly, several natural dietary supplements have undergone Phase II trials of cancer therapy; however, the majority failed to progress to Phase III trials, in spite of their positive results in pre-clinical investigations with animal models, cell lines, and/or small early phase clinical trials [4].

There are many challenges encountered in the development of herbal drugs. For example, randomized, placebo-controlled trials are crucial in the evaluation of any drug for health benefits or disease mitigation; however, the peculiar color, taste, and smell in herbal medicine make it difficult to conduct placebo-controlled trials. There were many occasions where the clinical evaluation of herbal drugs had shortcomings in trial design, improper execution, and weak data analysis, particularly due to the inappropriate number of patients in trials, improper randomization, and selection bias. Also conducting pharmacokinetic studies on polychemical natural products is quite complicated unless the active ingredient/active principle is known. The presence of several different active ingredients makes the pharmacokinetic evaluation more difficult and complex. Poor standardization and lack of quality control of herbal preparations as well as the presence of several active compounds would make the dose calculation a tedious process and sometimes lead to discrepancies in the dosage and treatment duration of the herbal remedies. Moreover, the contamination or adulterations of herbal preparations might result in undesirable toxic effects sometimes with dire consequences [4].

1.5 Conservation of medicinal plants

The use of herbal drugs shows an increasing global trend and the wild populations of plants having medicinal properties are facing many threats as a result. Therefore, the need for conservation of those plants has now become a priority [30, 43]. According to Heywood [30], this poses several problems. The numbers of medicinal plant species involved are large and information on the conservation status of the majority of the species are lacking. Threats faced by over-collecting, insufficient knowledge on the genetic variation and indigenous traditional knowledge are other problems. The reluctance of the policy makers to become involved and invest in conservation of these species is another critical issue.

1.6 The future trends in the medicinal plant research

Despite the numerous deterrents in the field of phytomedicine, researchers all over the world are conducting pre-clinical studies and clinical trials with botanicals to harness the maximum benefits from the healing powers of plants. Moreover, agencies like the National Institute of Health, USA; European Medicines Agency (EMA); Indian Council of Medical Research; and National Health and Medical Research Council, Australia, are undertaking clinical trials in assistance with several governments and private institutes [4].

The application of metabolomics in natural products research is a recent trend that is aimed at the qualitative and quantitative analysis of all the metabolites of an organism at a specific time and under specific conditions. Analytical techniques like high-resolution mass spectrometry and nuclear magnetic resonance spectroscopy are employed here to dereplicate and quantify the known metabolites against novel natural products. Along with metabolomics-guided fractionation tools, it is possible to identify active components at the first fractionation step, as well as to predict the metabolites that might be bioactive. Further, the metabolomics approach could help in the prioritization of fractions for further purification, saving time, and resources in isolating the target compounds [44].

With the progress in bioinformatics, computational techniques have entered in the process of drug discovery and development, and often precede or complement *in vitro* and *in vivo* studies. These computer-aided or *in silico* design is being utilized to expedite and facilitate hit identification, hit-to-lead selection, optimize the absorption, distribution, metabolism, excretion, and toxicity profile and avoid safety issues [45, 46]. These integrated computer-assisted strategies would be beneficial in processing a large amount of available structural and biological information within a short period of time for a straight-forward search of bioactive natural products [47].

Although there are significant numbers of very potent phytochemical compounds particularly with anti-tumor activity, the nonspecific administration (i.e. dosing the whole animal) during testing would lead to very high toxicities. The delivery of such agents specifically to the tumor area would enable to use those materials for treatment. In this respect, liposome-encapsulated toxic agents and antibody-conjugates with natural toxins/pure compounds would be an ideal strategy [22]. Moreover, the low bioavailability of phytochemicals would also hamper the further development of these agents. The use of nanoparticles is one of the promising strategies to significantly increase the bioavailability of natural products. The improvement in their pharmacokinetic properties might lead to a better therapeutic effect, without high-dose-induced acute toxicity. In this respect, polymeric nanoparticles have been employed to increase the bioavailability of luteolin, epigallocatechin gallate (EGCG), tea polyphenols, and silibinin while the oral bioavailability of apigenin was improved by incorporating it into a carbon nanopowder solid distribution [48]. Moreover, an improvement in the molecular targeting, oral bioavailability, and anticancer efficacy was observed for the new ginsenoside, 25-OCH3-PPD (GS25) isolated from *Panax notoginseng* upon its encapsulation into PEG-PLGA nanoparticles [49]. These examples indicate that the role of nanotechnology would be imperative to the field of herbal medicine in the coming years.

1.7 Conclusion

The plant-based healing systems continue to play an essential role in health care while functioning as an important source of novel pharmacologically active compounds. Despite the availability of compounds derived from computational and combinatorial chemistry as new drug leads as well as the challenges confronted over the years during the development of herbal-based drugs, phytochemicals still hold a fair share in drug discovery programs owing to their incomparable chemical diversity and novel mechanisms of action.

References

[1] Pan SY, Zhou SF, Gao SH, Yu ZL, Zhang SF, Tang MK, Sun JN, Ma DL, Han YF, Fong WF, Ko KM. New Perspectives on how to discover drugs from herbal medicines: CAM's outstanding contribution to modern therapeutics. Evid Based Complementary Altern Med 2013, 2013 Article ID 627375, 25 pages 10.1155/2013/627375.

[2] Sofowora A, Ogunbodede E, Onayade A. The role and place of medicinal plants in the strategies for disease prevention. Afr J Tradit Complement Altern Med 2013, 10(5), 210–229.

[3] Newman DJ, Cragg GM. Natural products as sources of new drugs from 1981 to 2014. J Nat Prod 2016, 79(3), 629–661.

[4] Chugh NA, Bali S, Koul A. Integration of botanicals in contemporary medicine: Road blocks, checkpoints and go-ahead signals. Integr Med Res 2018, 7(2), 109–125.

[5] Fitzgerald M, Heinrich M, Booker A. Medicinal plant analysis: A historical and regional discussion of emergent complex techniques. Front Pharmacol 2020, 10, 1480. 10.3389/fphar.2019.01480.

[6] Jamshidi-Kia F, Lorigooini Z, Amini-Khoei H. Medicinal plants: Past history and future perspective. J Herbmed Pharmacol 2018, 7(1), 1–7.

[7] Fabricant DS, Farnsworth NR. The value of plants used in traditional medicine for drug discovery. Environ Health Perspect 2001, 109(Suppl 1), 69–75.

[8] Yuan H, Ma Q, Ye L, Piao G. The Traditional Medicine and Modern Medicine from Natural Products. Molecules 2016, 21, 559. 10.3390/molecules21050559.

[9] Capasso L. 5300 years ago, the Ice Man used natural laxatives and antibiotics.Lancet 1998, 352(9143), 1864. 10.1016/S0140-6736(05)79939-6.

[10] Kelly K. the history of medicine early civilizations: Prehistoric times to 500 C.E. facts on file. NY, USA, 2009.

[11] Teall EK. Medicine and doctoring in ancient Mesopotamia. GVJH 2014, 3(1), Article 2 https://scholarworks.gvsu.edu/gvjh/vol3/iss1/2 Accessed on 10. 01.2020.

[12] Gu S, Pei J. Innovating Chinese herbal medicine: From Traditional health practice to scientific drug discovery.Front Pharmacol 2017, 16(8), 381. 10.3389/fphar.2017.00381.

[13] Petrovska BB. Historical review of medicinal plants' usage. Pharmacogn Rev 2012, 6(11), 1–5.

[14] Sewell RDE, Rafieian-Kopaei M. The history and ups and downs of herbal medicine usage. J HerbMed Pharmacol 2014, 3(1), 1–3.

[15] Khan H. Medicinal plants in light of history: Recognized therapeutic modality. J Evid Based Complementary Altern Med 2014, 19(3), 216–219.

[16] Kumar S, Dobos GJ, Rampp T. The significance of Ayurvedic medicinal plants. J Evid Based Complementary Altern Med 2017, 22(3), 494–501.

[17] Balkrishna A, Mishra RK, Srivastava A, Joshi B, Marde R, Prajapati UB. Ancient Indian rishi's (Sages) knowledge of botany and medicinal plants since Vedic period was much older than the period of Theophrastus, A case study- who was the actual father of botany?. Int J Unani Integ Med 2019, 3(3), 40–44.

[18] Halberstein RA. Medicinal plants: Historical and cross-cultural usage patterns. Ann Epidemiol 2005, 15(9), 686–699.

[19] Solomou AD, Martinos K, Skoufogianni E, Danalatos NG. Medicinal and aromatic plants diversity in Greece and their future prospects: A review. Agric Sci 2016, 4(1), 9–21.

[20] Scarborough J. Theophrastus on herbals and herbal remedies. J Hist Biol 1978, 11(2), 353–385.

[21] Hajar R. The air of history (Part II) medicine in the Middle Ages. Heart Views 2012, 13, 158–162.

[22] Newman DJ, Cragg GM, Snader KM. The influence of natural products upon drug discovery (Antiquity to late 1999). Nat Prod Rep 2000, 17(3), 215–234.

[23] Atanasov AG, Waltenberger B, Pferschy-Wenzig EM, Linder T, Wawrosch C, Uhrin P, Temml V, Wang L, Schwaiger S, Heiss EH, Rollinger JM, Schuster D, Breuss JM, Bochkov V, Mihovilovic MD, Kopp B, Bauer R, Dirsch VM, Stuppner H. Discovery and resupply of pharmacologically active plant-derived natural products: A review. Biotechnol Adv 2015, 33(8), 1582–1614.

[24] Saad B, Azaizeh H, Abu-Hijleh G, Said O. Safety of traditional Arab herbal medicine. Evid Based Complement Alternat Med 2006, 3(4), 433–439.

[25] Duarte DF. Opium and opioids: A brief history. Rev Bras Anestesiol 2005, 55(1), 135–146.

[26] Elliott B. The world of the Renaissance herbal. Renaissance Studies 2011, 25(1), 24–41.

[27] Classen A, Meyer FG, Trueblood EE, Heller JL. The Great Herbal of Leonhart Fuchs. De historia stirpium commentarii insignes.Ger Stud Rev 2001, 24(3), 595–597. 10.2307/1433419.

[28] Ghorbani A, Naghibi F, Mosaddegh M. Ethnobotany, ethnopharmacology and drug discovery. Iran J Pharm Sci 2006, 2(2), 109–118.

[29] Soejarto DD, Fong HHS, Tan GT, Zhang HJ, Ma CY, Franzblau SG, Gyllenhaal C, Riley MC, Kadushin MR, Pezzuto JM, Xuan LT, Hiep NT, Hung NV, Vu BM, Loc PK, Dac LX, Binh LT, Chien NQ, Hai NV, Bich TQ, Cuong NM, Southavong B, Sydara K, Bouamanivong S, Ly HM, Van Thuy T, Rose WC, Dietzman GR. Ethnobotany/ethnopharmacology and mass bioprospecting: Issues on intellectual property and benefit-sharing. J Ethnopharmacol 2005, 100(1–2), 15–22.

[30] Heywood VH. Biodiversity, conservation and ethnopharmacology. In: Heinrich M, Jaeger A eds. Ethnopharmacology. 1st Oxford, UK, John Wiley & Sons, 2015, 41–51.

[31] Pieroni A, Quave CL. Ethnobotany and biocultural diversities in the Balkans: Perspectives on sustainable rural development and reconciliation. NY, USA, Springer Science, 2014.

[32] Nolan JM, Turner NJ. Ethnobotany: The study of people-plant relationships in ethnobiology. In: Anderson EN, Pearsall D, Hunn E, Turner N eds. Ethnobiology. 1st NJ, USA, John Wiley & Sons, 2011, 133–145.

[33] Lozada M, Ladio AH, Weigandt M. Cultural transmission of ethnobotanical knowledge in a rural community of Northwestern Patagonia, Argentina. Econ Bot 2006, 60, 374–385.

[34] Reyes-García V. The relevance of traditional knowledge systems for ethnopharmacological research: Theoretical and methodological contributions. J Ethnobiol Ethnomed 2010, 6, 32. 10.1186/1746-4269-6-32.

[35] Napagoda MT, Sundarapperuma T, Fonseka D, Amarasiri S, Gunaratna P. Traditional uses of medicinal plants in Polonnaruwa district in North Central Province of Sri Lanka. Scientifica 2019, 2019, 9737302. 10.1155/2019/9737302.

[36] Khafagi IK, Dewedar A. The efficiency of random versus ethno-directed research in the evaluation of Sinai medicinal plants for bioactive compounds. J Ethnopharmacol 2000, 71, 365–376.

[37] Slish DF, Ueda H, Arvigo R, Balick MJ. Ethnobotany in the search for vasoactive herbal medicines. J Ethnopharmacol 1999, 66, 159–165.

[38] Albuquerque UP, Medeiros PMD, Ramos MA, Ferreira J, Washington S, Nascimento ALB, Avilez WMT, Melo JGD. 2014. Are ethnopharmacological surveys useful for the discovery and development of drugs from medicinal plants?. Rev Bras Farmacogn 2014, 24(2), 110–115.

[39] Clapp RA, Crook C. Drowning in the magic well: Shaman pharmaceuticals and the elusive value of traditional knowledge. J Environ Dev 2002, 11(1), 79–102.

[40] Cordell GA, Colvard MD. Some thoughts on the future of ethnopharmacology. J Ethnopharmacol 2005, 100(1–2), 5–14.

[41] Graham JG, Farnsworth NR. The NAPRALERT database as an aid for discovery of novel bioactive compounds. In: Mander L, Liu HW, eds, Comprehensive natural products II. 1st, Amsterdam, The Netherlands, Elsevier Science, 2010, 81–94.

[42] Honek J. Preclinical research in drug development. Medical Writ 2017, 26(4), 6–8.
[43] Randriamiharisoa M, Kuhlman AR, Jeannoda V, Rabarison H, Rakotoarivelo N, Randrianarivony T, Raktoarivony F, Randrianasolo A, Bussmann RW. Medicinal plants sold in the markets of Antananarivo, Madagascar. J Ethnobiol Ethnomed 2015, 11, 60. 10.1186/s13002-015-0046-y.
[44] Harvey AL, Edrada-Ebel R, Quinn RJ. The re-emergence of natural products for drug discovery in the genomics era. Nat Rev Drug Discov 2015, 14(2), 111–129.
[45] Harvey AL. Natural products in drug discovery. Drug Discov Today 2008, 13(19–20), 894–901.
[46] Kapetanovic IM. Computer-aided drug discovery and development (CADDD): In silico-chemico -biological approach. Chem Biol Interact 2008, 171(2), 165–176.
[47] Rollinger J, Langer T, Stuppner H. Integrated in silico tools for exploiting the natural products' bioactivity. Planta Med 2006, 72(8), 671–678.
[48] Watkins R, Wu L, Zhang C, Davis RM, Xu B. Natural product-based nanomedicine: Recent advances and issues. Int J Nanomedicine 2015, 10, 6055–6074.
[49] Voruganti S, Qin JJ, Sarkar S, Nag S, Walbi IA, Wang S, Zhao Y, Wang W, Zhang R. Oral nano-delivery of anticancer ginsenoside 25-OCH3-PPD, a natural inhibitor of the MDM2 oncogene: Nanoparticle preparation, characterization, *in vitro* and *in vivo* anti-prostate cancer activity, and mechanisms of action. Oncotarget 2015, 6(25), 21379–21394.

Thushara Diyabalanage

2 Plant secondary metabolites as prospective pharmaceuticals and cosmeceuticals

2.1 Primary metabolites and secondary metabolites

Primary metabolites can be defined as the molecules that are directly involved in the conduction of basic life functions such as cell division, growth, respiration, storage, and reproduction. The nutritional components carbohydrates, lipids, amino acids, and alcohols belong to this category [1]. Generally, they are simpler in structure and widely distributed in all parts of the plants.

However, identifying the role of secondary metabolites and their role in an organism needs a lot more explanation. Unlike the primary metabolites, the secondary metabolites of plants are present in low abundance and stored in dedicated cells or organs. Structurally they demonstrate more complexity, diversity, and uniqueness. The definition of secondary metabolism itself has been subjected to intense debate for many years. Sometime back, secondary metabolites were identified as a group of molecules that do not play a specific role in an organism. Nevertheless, based on more recent interpretations, they are defined as a group of molecules that afford some distinct advantage to the organism to be fitter during the natural selection [2, 3].

Secondary metabolites play a key role in a plant's adaptation to the environment. They contribute to the plant's fitness by performing the roles of antibiotics, antifungal, and antiviral agents preventing the infections and attacks of pathogens. Secondary metabolites, with the ability of UV absorption, can protect against leaf damage by UV radiation in intense sunlight. Similarly, secondary metabolites can have numerous attractive biological properties such as antifeedant, insecticidal, and larvicidal properties which would support the organism by providing protection against predation, parasitism, invasion, and competition [3].

Plant secondary metabolites can be classified into several major groups based on their chemistry and biosynthetic pathways as phenolics, terpenoids, steroids, and alkaloids [4]. Phenolics are a large family of secondary metabolites ubiquitously distributed among the higher plants. They can be subdivided into several subgroups based on chemistry such as flavonoids, lignans and polyphenols, coumarins, and anthocyanins. Similarly, terpenoids are also widespread in the plant kingdom. They can be further divided into subgroups based on biosynthesis, considering the number of

Thushara Diyabalanage, Actives International LLC, 6 Pearl Court, Unit G, Allendale, NJ 07401, USA, e-mail: thusharad@activesinternational.com

https://doi.org/10.1515/9783110595949-002

isoprene units involved to build the terpene scaffold. The distribution of alkaloids in plants is sparse and more specific to specific plant genera and species.

2.2 Plant secondary metabolites as a source for pharmaceuticals

These plant secondary metabolites are enriched with an extraordinary array of attractive biological activities. Their potential to cure various ailments has been known and explored since early civilizations [4, 5]. As a result, the plants had been a generous and excellent source of medicinal ingredients for many indigenous medicinal systems in various parts of the world. With the advancement of analytical chemistry and biology, man's quest to discover pure bioactive molecules responsible for those attractive medicinal properties from such herbal preparations used in folk medicine led to the development of many pharmaceuticals. With the rapid development of the pharmaceutical industry, some of them eventually have even become blockbuster drugs [4, 5]. A survey conducted on plant-derived pure compounds used as drugs indicated that 122 compounds identified were derived from 94 plant species. Interestingly, 80% of these drugs were clinically applied for the same purpose they were used for in folk medicine, further highlighting the significance of the ethnobotanical impact on natural product–based drug discovery [5].

These first generation of pharmaceuticals were developed by the observation of their efficacy in indigenous medicinal systems and targeted isolation of the pharmacologically active natural products from the native plant extract [6]. Eventually, some of these drug molecules were synthesized via more efficient routes and the synthetic product was favored over the natural product as it was commercially more feasible. The direct synthesis of the natural product addressed problems associated with plant collections and supply. During the structure–activity relationship studies of a natural product, a large number of synthetic analogs were derivatized. In some instances, some of these semi-synthetic derivatives displayed better biological activity and solubility so that they were chosen as better drugs. Bioactivity studies on natural products might reveal certain pharmacophores that can interact with some substrates with a therapeutic effect. Medicinal chemists have been able to design and develop synthetic drug molecules that incorporated such pharmacophores.

Therefore, natural product–based pharmaceuticals can be categorized as follows.
1. Pharmaceuticals developed by isolating the pure natural product from its native extract.
2. Synthetic product identical to the natural product.
3. Semi-synthetic derivative of the pure natural product to enhance efficacy and delivery.
4. Products of total synthesis, but the pharmacophore was inspired or based on a natural product.

2.3 First generation of natural products–based pharmaceuticals

Morphine (Figure 2.1), a phenanthrene-type alkaloid derived from *Papaver somniferum*, is the first plant-derived pure natural product to be commercially used as a therapeutic. It was marketed by Merck in 1826 as a sedative [7]. Based on the analgesic properties observed in salicin, a phenolic glucoside found in the bark of *Salix alba* (Willow), aspirin was developed by Bayer in 1899. Piria, an Italian chemist, resolved the chemical structure of salicin and converted it to salicylaldehyde in 1839. Aspirin (acetylsalicylic acid) was eventually synthesized by Gerhard in 1853. Subsequently, several other plant-derived drugs came into the market. This included other opium-derived alkaloids codeine (Figure 2.1) and papaverine (Figure 2.2) [4]. Cardiac glycoside digitoxin (Figure 2.3) extracted from the plant *Digitalis purpurea* (foxglove) was used to treat certain cardiac conditions [8]. Even though digitoxin is rarely used in current clinical applications, digoxin (Figure 2.3), another cardenolide, with a very close structure and isolated from the same plant, is still in use for various cardiac complications.

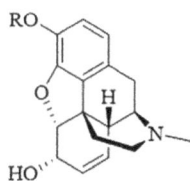

Morphine R = H
Codeine R = CH₃

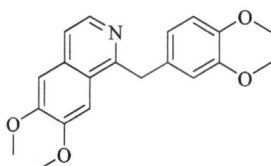

Figure 2.1

Papaverine

Figure 2.2

2.3.1 Anti-malaria drugs

Malaria has been a major global health challenge for many decades. The first anti-malaria drug quinine (Figure 2.4) was isolated from the bark of *Cinchona officinalis* by French pharmacists Caventou and Pelletier in 1920 [5, 9]. It was a classic story of drug discovery based on folkloric use. Indigenous tribes in the Amazon region of South America had been using the bark of this plant to treat fever successfully and that seemed to have prompted people in Europe to use it to treat malaria around the 1700s. The outstanding results of this treatment inspired Caventou and Palletier to isolate the active ingredient quinine, an alkaloid belonging to the quinolone group. Quinine provided the basic structural motif for the development of more effective malaria drugs such as chloroquine (Figure 2.5) and mefloquine that were synthetic modifications of quinine [5, 9].

With the *Plasmodium* parasite developing resistance over quinine-based drugs, they are no longer used to treat malaria. However, some of these quinine analogs have been very successfully employed to treat arthritis and some debilitating auto-immune diseases such as lupus and Sjogren's syndrome as disease-modifying drugs [10].

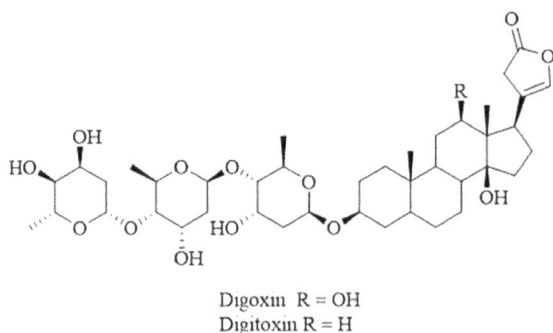 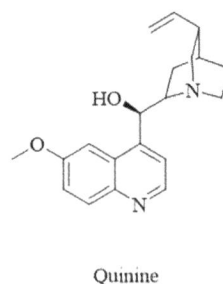

Digoxin R = OH
Digitoxin R = H

Figure 2.3

Quinine

Figure 2.4

Artemisinin (Figure 2.6), a sesquiterpenoid lactone, is another plant-based drug against malaria that has been developed by the observation of its efficacy in indige-nous medicine [9]. In ancient Chinese medicine, the leaves of the plant Quing-hao had been successfully used to treat malaria for about 2,000 years. The active com-pound artemisinin was isolated from *Artemisia annua* (Asteraceae) around 1970 and later on reported by the researchers at the Walter Reed Army Institute of Re-search in the USA in 1984 [5]. Eventually, a large number of artemisinin analogs have been synthesized and they have shown much better efficacy against malaria. The current treatment of malaria involves a combination therapy based on artemisi-nin and other drugs [9].

2.3.2 Anti-cancer drugs

– Vinca alkaloids

Vinca alkaloids, vinblastine, and vincristine (Figure 2.7) are some of the first anti-cancer drugs to have been developed from plants. They were isolated from Mada-gascar periwinkle, (*Catharanthus roseus*), a plant that the natives of Madagascar used to treat diabetes [5, 11]. The Madagascar periwinkle plant contained many ter-pene-indole alkaloids and out of them, vinblastine and vincristine showed the abil-ity to control the increase of lymphocytes. This was due to their ability to inhibit the cell division by binding to tubulin and thereby preventing the formation of the spindle. Vinblastine is used to treat Hodgkin's lymphoma, advanced testicular cancer, and breast cancer whereas vincristine is used to treat acute leukemia and other lymphomas [11, 12].

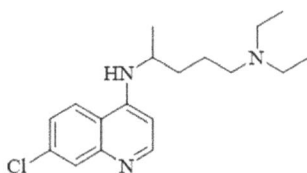

Chloroquine

Figure 2.5

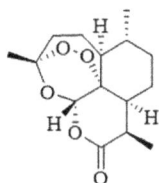

Artemisinin

Figure 2.6

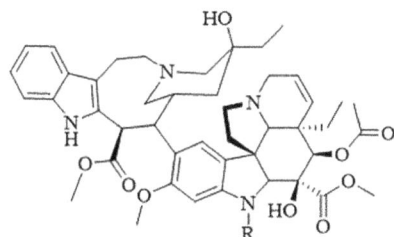

Vinblastine   R = CH₃

Vincristine   R = CHO

Figure 2.7

– **Taxanes**

Taxol (Paclitaxel) (Figure 2.8) is one of the most potent and widely used anticancer drugs discovered from plants. It is extensively used to treat breast, lung ovarian, bladder, prostate, and many other forms of cancers. The honor of the discovery of Taxol from the bark of the plant *Taxus brevifolia* (Pacific Yew tree) in 1966 goes to Mankush Wani and Monroe Wall of Research Triangle Institute in North Carolina [5, 13]. Its mode of action was deduced as stabilization of microtubules leading to the inhibition of spindle formation by Schiff and Susan Horwitz in 1980 [14]. Taxol was clinically approved to treat ovarian cancer in 1992 and breast cancer in 1994 and has become a blockbuster drug. Since then, several clinically more efficacious and deliverable forms of Taxol have been developed incorporating different advanced technologies such as nanotechnology and polymer science [5, 12]. The discovery of Taxol brought great recognition toward natural product–based drug discovery programs. It gave rise to a renaissance of research activity to explore new therapeutic leads from hitherto uninvestigated higher plants.

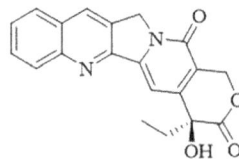

– **Camptothecin**
Camptothecin (Figure 2.9) is a pentacyclic alkaloid isolated from the plant *Camptotheca acuminata*, a Chinese medicinal plant widely used in cancer treatment in traditional medicine [15]. The discovery of camptothecin is also credited to Makush Wani and Monroe Wall of Research Triangle Institute in North Carolina, who isolated this compound during a screening campaign of medicinal plant extracts. Its mechanism of action has been determined as the inhibition of DNA topoisomerase-I that eventually led to the arrest of the cell cycle at the S-phase [15, 16]. A major challenge in developing a cancer drug from camptothecin was its low solubility [4]. This has been addressed by several synthetic modifications [16]. As a result, there are three semi-synthetic derivatives of camptothecin—topotecan (Figure 2.10), irinotecan (Figure 2.11), and belotecan—isolated from *C. acuminata* that are used in clinics as chemotherapeutic agents to treat cancer [17].

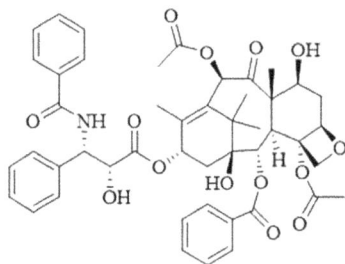

Paclitaxel

Figure 2.8

Camptothecin

Figure 2.9

– **Podophyllotoxins**
Podophyllotoxins have been used as folk medicine in numerous cultures for many centuries. Their subsequent clinical evaluation confirmed an anticancer activity. Chemically, they belong to the group of lignans and are formed by the addition of two cinnamic acid moieties. However, their toxicity prevented them from being developed for clinical use [18]. An attempt to develop semi-synthetic derivatives of some podophyllotoxins that retain their anticancer activity with less toxicity from Indian *Podophyllum* root extract led to the discovery of two anticancer drug leads that displayed enhanced antineoplastic activity. Their synthesis gave two drugs etoposide and teniposide (Figure 2.12) [17, 19]. The mechanism of action of these two

has been elucidated as DNA topoisomerase-II inhibition, arresting the cell cycle at the G2-phase, thus preventing the replication of DNA [17].

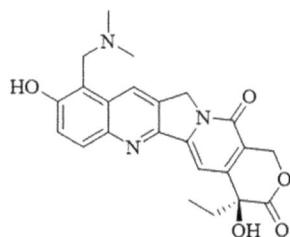

Topotecan

Figure 2.10

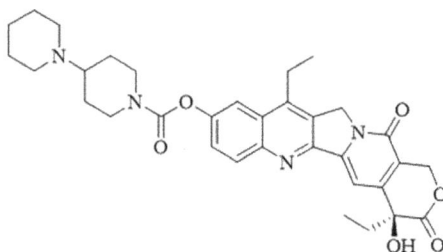

Irinotecan

Figure 2.11

– **Combretastatins**

The combretastatins (Figure 2.13) are a group of stilbenes, isolated from the South African "bushwillow" *Combretum caffrum* collected in the Southeastern region of Africa in the 1970s. Combretastatins have a mechanism of action similar to that of Taxanes and Vinca alkaloids by binding to β-tubulin and thereby inhibiting tubulin polymerization. They also act as antiangiogenic agents, by selectively preventing the blood flow in tumors resulting in vascular shutdown and subsequent tumor necrosis [16, 17]. Poor solubility of combretastatin A-4 (Figure 2.13) in an aqueous media was a significant drawback toward its further development as a drug candidate. The water-soluble analog, combretastatin A-4 phosphate has overcome that hurdle and has received orphan drug status from the US Food and Drug Administration (FDA) for the treatment of a range of thyroid cancers and ovarian cancers. It is in advanced clinical trials against anaplastic thyroid cancer, in combination with paclitaxel [17].

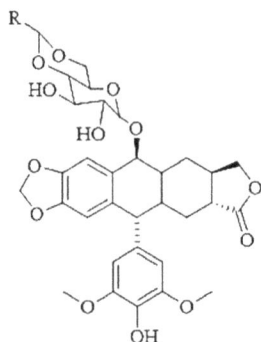

Etoposide   R = CH₃
Teniposide   R =

Figure 2.12

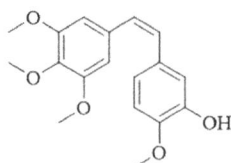

Combretastatin A-4 (**17**)

Figure 2.13

– **Homoharringtonine**

Homoharringtonine is a cytotoxic alkaloid isolated from *Cephalotaxus harringtonia* used in traditional Chinese medicine [16]. Its mode of action has been elucidated as blocking the synthesis of peptidyl transferase center, resulting in apoptosis [12, 17]. After a 40-year-long wait to get the FDA approval, finally in 2013 a semi-synthetic homoharringtonine, named omacetaxine, was approved to treat chronic myeloid leukemia. It has been observed that the usage of homoharringtonine made it possible to achieve remission in 92% of the patients [12, 17].

2.3.3 Other drugs

Some plant-based natural products used in traditional folk medicine for certain disease conditions have inspired the development of several well-established drugs. *Ammi visnaga*, a plant grown in the Mediterranean region and Egypt, has been used in folk medicine to treat a variety of ailments. The active ingredient present in this plant is identified as khellin (Figure 2.14), a smooth muscle relaxant that has several side effects. Sodium cromoglycate (Figure 2.15) a synthetic derivative of khellin, has shown fewer side effects and has been successfully used as a bronchodilator. Thus, the development of cromolyn (sodium cromoglycate) was directly influenced by the natural product khellin [5].

Galega officinalis, better known as Goat's rue or Italian lilac, is a plant native to the Middle East that has been extensively used to treat diabetes since the Middle Ages. Chemical investigations of the plant extracts traced the antidiabetic activity to a group of guanidine derivatives, including galegine and isoprenyl guanidine that showed the least toxicity. Further studies involving galegine (Figure 2.16) inspired the development of the pharmacophore that led to the synthesis of metformin (Figure 2.17), a drug well established to treat type II diabetes [4, 5].

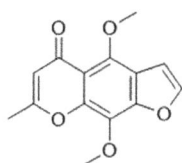
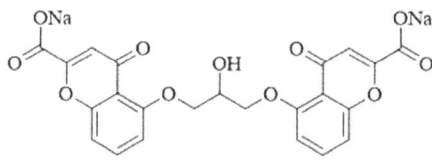
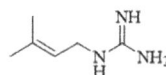

Kehellin	Chromoglycate	Galegine
Figure 2.14	**Figure 2.15**	**Figure 2.16**

Pilocarpine extracted from the leaves of the South American plant *Pilocarpus microphyllus* has been long used as a treatment to regulate the eye pressure associated with glaucoma. Its ability to stimulate the production of saliva in salivary glands has prompted its use to treat xerostomia associated with Sjogren's syndrome [8].

2.4 Recent advances

Plants still continue to be a rich source for the discovery of pharmaceuticals. There are hundreds of plant-derived drugs in clinical trials [4, 5, 16, 20]. With recent advances in biotechnology and analytical chemistry, scientists are better equipped in their search for new bioactive molecules using a myriad of powerful tools [4]. Rapid high-throughput screening methods aided by robotics have facilitated the screening of a large number of plant extracts and fraction of samples in a shorter time. Advanced de-replication methods have enabled the identification of active metabolites more efficiently. It is estimated that only about 6% of the higher plants of nearly 300,000 species have been subjected to a systematic investigation [16]. Nevertheless, there is a growing body of evidence to suggest that some portion of the secondary metabolites that we have identified as plant metabolites may not have true plant origins and they can be biosynthesized by the endophytes [4].

Meanwhile, the dietary supplement industry that uses standardized plant extracts containing natural products seems to have thrived in the last few years [4, 21]. The long time taken for a plant-derived drug to get the FDA approval and more stringent regulatory requirements have apparently promoted the attention toward botanicals which have relatively lesser regulations to reach the consumers. Thus, there is a significant steadily growing demand for the dietary supplements that address a vast array of health claims. Recently, the FDA approved the first botanical drug, an ointment containing sinecatechins to be used against genital warts due to human papillomavirus (HPV) infection [22]. This drug contains a proprietary blend of eight green tea catechins and other green tea components from partially purified *Camellia sinensis* extract.

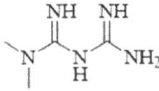

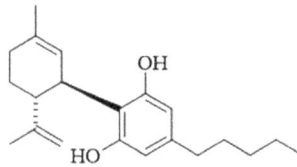

Metformin

Canabidiol

Figure 2.17   **Figure 2.18**

Similarly, there had been a hive of activity following the discovery of cannabinoid receptors and their significance in the development of treatments against a variety of debilitating disease conditions. In 2018, FDA approved epidiolex, an oral solution of cannabidiol (Figure 2.18) derived from marijuana for the treatment of seizures associated with two rare, but more severe, forms of epilepsy [23]. The FDA also has approved two other synthetic cannabinoids, dronabinol and nabilone, to treat nausea associated with chemotherapy. Cannabis-based treatments have already shown tremendous promise in the treatment of several neurological disorders. Several such drug candidates are in clinical trials against multiple sclerosis, Parkinson's disease, and Alzheimer's disease

2.5 Cosmetic ingredients and cosmeceuticals

A cosmetic can be defined as a product that can cleanse, beautify, and promote attraction after topical application [24]. It is not supposed to alter the structure of the skin, nor heal it and does not require to undergo a stringent approval process for their label claims, despite the requirement to adhere to safety regulations. The word "cosmeceutical" was first introduced by Albert Kligman in 1984 to describe a substance that exerts both cosmetic and therapeutic benefits [24, 25]. A cosmeceutical is a biologically active ingredient that can result in a cosmetic effect after application and it may change the skin structure like a pharmaceutical while achieving its objective. Thus, cosmeceuticals can be introduced as cosmetic-pharmaceutical hybrids used to maintain and improve the appearance and beauty of the skin [26]. The FDA of the USA does not recognize the word "cosmeceuticals." According to the Federal Food Drug and Cosmetic Act, any ingredient that claims drug-like properties needs to undergo an extensive review process to substantiate those claims and ensure safety before approval for the use of public.

A large number of plant-derived secondary metabolites are used as cosmeceuticals. They are used to improve and nourish skin appearance. The skin being the largest organ in the human body provides protection against external elements and functions as a very efficient barrier that protects the internal environment. Various secondary metabolites present in cosmeceuticals play very important roles in protecting the integrity of the skin and keeping it healthy. Based on their specific roles they can be categorized into several groups as follows. During formulation of the cosmetic products, these secondary metabolites are used in pure forms or as botanical extracts where active ingredients may be present in lower percentages.

(i)   Antioxidants
(ii)   Anti-inflammatory agents
(iii)   Skin lighteners
(iv)   MMP (matrix metalloproteinase) inhibitors
(v)   Elastase inhibitors
(vi)   Antiangiogenics
(vii)   Other compounds

2.5.1 Antioxidants

The skin being the outermost layer of the body gets regularly exposed to UV radiation, which triggers the rapid generation and accumulation of reactive oxygenated species (ROS) [27]. Generally, ROS are generated by regular cell metabolism as well. The presence of an endogenous antioxidant system actually maintains the balance. However, when the equilibrium is shifted toward more ROS, the pathological process of oxidative stress occurs resulting in cell damage inducing photo-aging [27]. When this endogenous antioxidant defense weakens, due to the action of ROS, damages can occur to proteins, enzymes, lipids, and DNA.

In order to support the endogenous antioxidants of the body, exogenous antioxidants can be used very successfully. This includes the antioxidants supplied by food and antioxidants that can be applied topically in the form of lotions and creams to provide protection. Thus a large number of plant-based phenolic compounds that can act as potent antioxidants have been successfully incorporated into cosmetic formulations to reduce the impact of the photo-damage [28]. These include flavonoids (catechins, isoflavones, proanthocyanidins, and anthocyanins), phenolic acids (benzoic, gallic, cinnamic), and stilbenes from plants such as grapes, tea, grapefruits, and oranges. Such plant-based secondary metabolites protect the skin from photo-aging by absorbing UV radiation and inhibiting UV-induced free radical reactions in the cells. They also modulate the endogenous antioxidant and inflammatory systems [26].

2.5.1.1 Plant-derived antioxidants used as cosmeceuticals

– **Resveratrol**

Grape seed and skin are rich in a variety of polyphenols such as quercetin, catechin, gallic acid, and proanthocyanidins that contribute to skin protection as antioxidants and anti-inflammatory agents. One of the principal constituents in grape seed extract, resveratrol (Figure 2.19), is a stilbene which is a potent antioxidant and anti-inflammatory agent as it could modulate many inflammatory cytokines such as IL-6, IL-8, and TNF-α. Thus, the topical application of grape seed extract has proven to reduce the UVB-induced oxidative damage and inflammation in human keratinocytes [26].

– **Carotenoids**

Carotenoids function as accessory light-harvesting pigments during photosynthesis. However, they also have the ability to prevent photooxidative damage. β-carotene, lycopene (Figure 2.20) from tomatoes, canthaxanthin, and lutein (Figure 2.21) are some common carotenoids that have been incorporated in cosmetic formulations [27, 28].

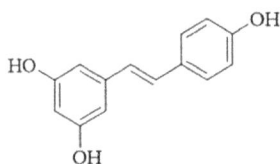

Resveratrol

Figure 2.19

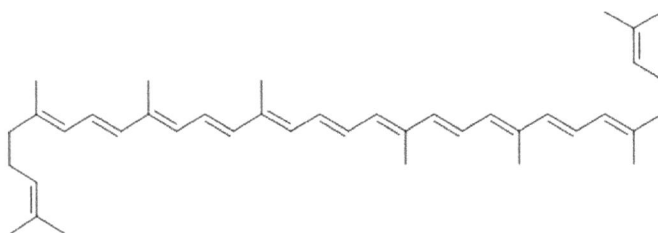

Lycopene

Figure 2.20

– **Apigenin**

Apigenin (Figure 2.22) is a unique flavone with multiple beneficial properties toward the skin. It is a potent antioxidant present in some citrus fruits including grapefruit [27]. It also displays potent anti-inflammatory activity with the ability to inhibit PLA-2. Hence it is very effectively used as a cosmeceutical.

2.5.2 Anti-inflammatory agents

Many plant-derived secondary metabolites employed as cosmeceuticals function as potent anti-inflammatory agents as they have the ability to modulate several biochemical pathways involved in the inflammatory cascade. Asiatic acid (Figure 2.23) and madecassic acid (Figure 2.24), two triterpenoids present in *Centella asiatica*, can inhibit enzymes COX-2, iNOS, and pro-inflammatory cytokines IL-6, 1 L-1β, and TNF-α [29]. 18-β Glycyrrhetinic acid (Figure 2.25), a triterpenoid extracted from licorice root, is another potent anti-inflammatory agent. It inhibits the activation of NF-κB and the activities of phosphoinositide-3-kinase (PI3K) p110δ and p110γ isoforms and reduces the production of lipopolysaccharide-induced tumor necrosis factor-α (TNF-α), interleukin (IL)-6, and IL-1β [30]. Curcumin (Figure 2.26), isolated from Curcuma longa, is another phenolic compound that has strong anti-inflammatory activity, with the ability to inhibit COX 2 and NF-κB [31]. All of these secondary metabolites are extensively used as cosmeceuticals. The deep yellow color of curcumin extract is undesirable for the cosmetic industry. Therefore, tertrahydrocurcumin, which has a much more neutral color, has been used.

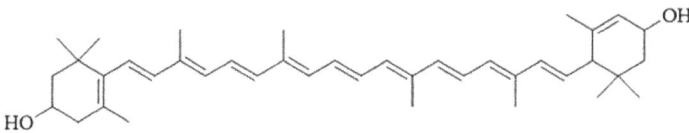

Lutein

Figure 2.21

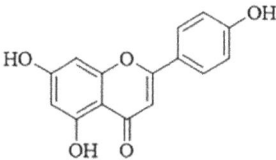

Apigenin

Figure 2.22

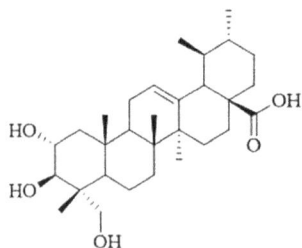

Asiatic acid

Figure 2.23

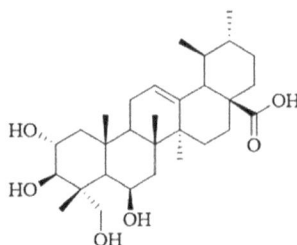

Madecassic acid

Figure 2.24

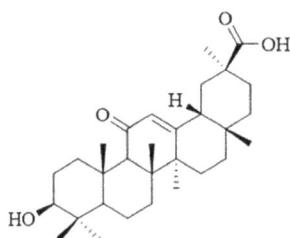

18β-Glyceyrrhetenic acid

Figure 2.25

2.5.3 Skin lighteners

The tan color of the human skin is due to the presence of the dark-colored pigment, melanin. It performs the crucial role of protecting the skin from UV radiation. In addition to that, the level of melanin present in the skin determines the various shades of phenotypic appearances of skin color [32]. The synthesis of melanin occurs in melanocytes located in the basal layer of the epidermis. It is subsequently distributed among the keratinocytes via melanosomes. Melanin is a bio-polymer formed by polymerization of tyrosine units [33]. The enzyme tyrosinase is involved in this process of melanogenesis and thus the inhibitors of tyrosinase can be used for skin lightening. The abnormal accumulation of melanin in the skin or irregular hyperpigmentation can be an aesthetic problem. Compounds that have hypopigmentation activity are used in cosmetics and dermatology to address this. Most of the hypopigmentation agents are tyrosinase inhibitors and they can be divided further based on their mode of action [32].

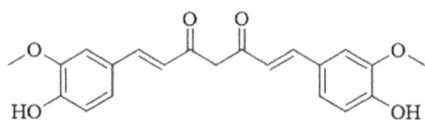

Curcumin

Figure 2.26

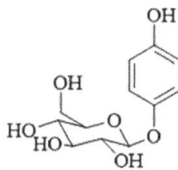

Arbutin

Figure 2.27

- **Arbutin**

Arbutin (Figure 2.27), β-D glucopyranoside of hydroquinone is well known for its ability to inhibit the activity of tyrosinase [34]. Therefore, it is widely used as a skin lightener. Natural arbutin is obtained from the leaves of *Vaccinicum vitis-idaea* and related plants [33]. However, in current cosmetic applications, the synthetically produced arbutin is used over the natural product due to the higher cost for the production of natural arbutin.

- **Glabridin**

Glabridin (Figure 2.28), a prenylated isoflavonoid extracted from the root of perennial herb *Glycyrrhiza glabra* (Licorice) is a potent skin lightener. It is known to contribute to skin lightening in two different routes. Being a potent antioxidant, glabridin has the ability to scavenge free radicals, thus it inhibits UVB-induced pigmentation. Glabridin also inhibits tyrosinase without affecting DNA synthesis [32, 33]. It has good anti-inflammatory properties as well. It has been found out that the *in vitro* skin-lightening effect of glabridin is 16 times greater than that of hydroquinone.

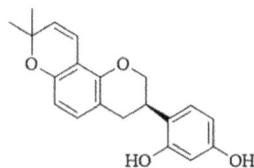

Glabridin

Figure 2.28

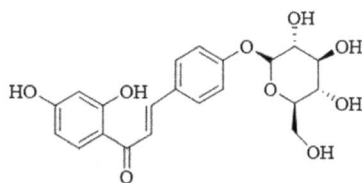

Isoliquiritin

Figure 2.29

- **Liquiritin and isoliquiritin**

Liquiritin and isoliquiritin (Figure 2.29) are two other flavonoids present in licorice that show skin-lightening properties. These compounds can disperse melanin in addition to antioxidant and anti-inflammatory properties [34].

- **Ellagic acid**

Ellagic acid (Figure 2.30), a polyphenol present in a variety of fruits and nuts, has shown attractive skin-lightening properties. It has displayed the ability to inhibit melanogenesis by reducing tyrosinase activity. Due to the polyphenolic nature, ellagic acid too can scavenge free radicals and reduce the UVB-induced hyperpigmentation [34].

2.5.4 MMP inhibitors

Matrix metalloproteinases (MMPs) are a group of proteolytic enzymes that are involved in the degradation of extracellular matrix (ECM) proteins such as collagen, elastin, fibronectin, and proteoglycans. Prolonged exposure to UV radiation increases the expression of MMPs in human skin [35]. These MMPs play a vital role in photo-aging and even in photocarcinogenesis. Clinical signs for photo-aging are characterized by coarse wrinkles, blotchy dyspigmentation, telangiectasia, sallowness, increased fragility, and rough skin texture as the MMPs degrade the ECM proteins such as collagen and elastin that provide structural and functional support to the skin tissue [36]. Thus, MMP inhibitors are considered promising targets to combat photo-aging. Several plant-derived secondary metabolites have shown the ability to inhibit MMPs.

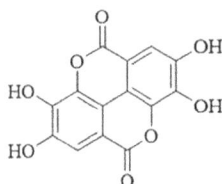

Ellagic acid

Figure 2.30

- **α-Mangostin**

α-Mangostin (Figure 2.31), the principal xanthone present in the rind or pericarp of *Garcinia mangostana*, significantly decreases the expression of multiple matrix-degrading proteinases, including matrix metalloproteinase-2(MMP-2), matrix metalloproteinase-9 (MMP-9). It also acts as a potent antioxidant. Therefore, α-mangostin is used in many skincare preparations to arrest the photo-damage [37].

- **Ellagic acid and its analogs**

The bark extract of *Anogeissus leiocarpus* shows potent inhibition against several MMPs. It shows inhibition against MMP 2, MMP 3, MMP 7, MMP 9, MMP 10, MMP-12, and MMP 14. This activity has been traced to ellagic acid and its derivatives [38]. Pomegranate seed extract, which is enriched with a higher content of ellagic acid, also showed inhibition of MMP-2 and MMP-3. In addition to that, the presence of these polyphenols that serve as strong antioxidants, provides photoprotection [39].

– **Apigenin**
Apigenin, a flavonoid present in the peel of certain citrus fruits such as grapefruit, displays significant MMP inhibition. It can reduce the expression of multiple MMP genes and up-regulate the expression of an inhibitor of MMP [40, 41]. Also, it has shown attractive anti-inflammatory activity by the inhibition of phospholipase A2.

2.5.5 Elastase inhibitors

Elastase is a proteolytic enzyme belonging to the serine protease group that can break down the extracellular matrix. Following an injury to the skin, neutrophils are recruited to the site, by the cytokines IL-9 and TNF-α. These neutrophils secrete the proteolytic enzyme elastase that degrades elastin, an important constituent of the extracellular matrix that maintains the firmness and elasticity of the skin [42]. Inhibiting the release and activity of elastase helps prevent tissue damage that can lead to loss of firmness and other signs of ageing. Therefore, plant-derived inhibitors of elastase are of extreme importance [43].

– **Boswellic acid**
Boswellic acids are triterpenoids present in the gum resin of the bark of the *Boswellia serrata* plant and act as potent inhibitors of the enzyme elastase [44]. In addition to that, they can protect the skin in a variety of other mechanisms. They have shown the ability to inhibit the formation of leukotriene B4, a pro-inflammatory mediator, and this prevents inflammation [45, 46]. Cathepsin D is a proteolytic enzyme that degrades desmosomes during epidermal desquamation. The inhibitors of cathepsin D can help to maintain the integrity of the stratum corneum and improve barrier function and may help to improve psoriatic skin. Boswellia extract shows the ability to inhibit cathepsin D [47]. *Boswellia* extract is enriched with six isomeric forms of boswellic acids. 11-Keto-β-boswellic acid and 3-O-acetyl, 11-keto-β-boswellic acid (Figure 2.32) are the two main *Boswellia* acids in this group when the yields are concerned.

2.5.6 Anti-angiogenics

Many conditions such as prolonged UV exposure and chronic inflammatory diseases like psoriasis and rosacea can give rise to undesired capillary growth or angiogenesis. Eventually, angiogenesis can lead to spider vein formation or telangiectasia. It has been found that some plant-based secondary metabolites can reduce inflammation and angiogenesis.

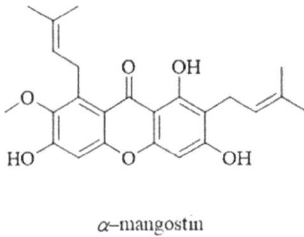

α–mangostin

Figure 2.31

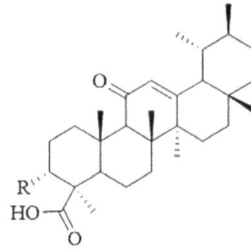

3-hydroxy, 11-keto-β–boswellic acid R = OH
3-O-acetyl, 11-keto-β–boswellic acid R = OAc

Figure 2.32

– **Humulone**

The cones of hops (*Humulus lupulus*) used in the beer industry contain a variety of compounds with important bioactivities [48]. Humulones are a group of prenylated flavones present in hops that are considered to be associated with the bitter taste. These humulones have potent anti-inflammatory, anti-microbial, and anti-angiogenic properties. Chemically, they belong to the category of alpha acids [48, 49]. The major drawback associated with these humulones is their relative instability. To address this in beer manufacturing, they are converted to corresponding iso-alpha acids which are more stable, via an isomerization reaction. Iso-alpha acids such as adhumulone (Figure 2.33) are used as anti-angiogenic agents in skincare [50].

– **Epigallocatechins**

Epigallacatchin-3-gallate (EGCG) (Figure 2.34) is the main polyphenol present in green tea. It has numerous important biological activities such as antioxidant, anti-inflammatory, immuno-modulatory, and anti-angiogenic activity. The green tea extract containing EGCG can prevent the induction of HIF-α (hypoxia inducible factor) and the expression of VEGF (vascular endothelial growth factor) [51]. Therefore, the topical application of green tea extract enables the prevention of angiogenesis.

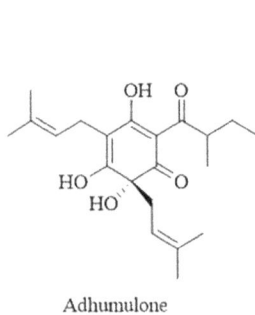

Adhumulone

Figure 2.33

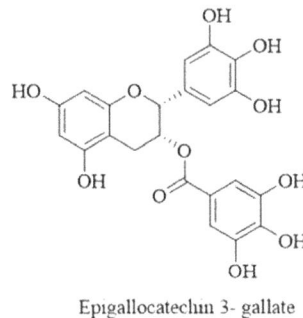

Epigallocatechin 3- gallate

Figure 2.34

2.5.7 Other compounds

- **Forskolin**

Forskolin (Figure 2.35), a diterpenoids present in the root of the South Asian medicinal plant *Coleus forskohlii*, can provide protection against UV damage in multiple roles [52]. It has the unique ability to directly activate adenyl cyclase to increase the production of cyclic AMP, which protect keratinocytes from UVB-induced apoptosis independent of the melanin present in the skin. Forskolin also promotes DNA repair by the removal of the UV damaged-pyrimidine dimers and phyto products [53].

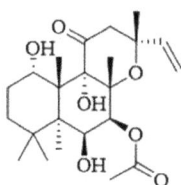

Forskolin

Figure 2.35

2.6 Concluding remarks

It is intriguing to see that plants have biosynthesized a vast array of molecules with bewildering structural diversity and unique bioactivity as an adaptation toward the environment. These plant-based secondary metabolites have given rise to a steady stream of drug molecules to successfully combat several killer diseases. Such natural product–derived pharmaceuticals and those developed using the pharmacophores-based on the observed structure–activity relationships of natural products have immensely contributed toward saving millions of lives. Plants still continue to be an exceptional source for the discovery of the next generation of pharmaceuticals.

Similarly, a large number of plant-derived cosmeceuticals are included in modern-day cosmetic formulas. These natural products address a myriad of problems associated with the skin, with their unique biological properties to make it look healthy and youthful.

References

[1]   Firn RD, Jones CG. The evolution of secondary metabolism – a unifying model. Mol Microbiol 2000, 37(5), 989–994.

[2]   Bourgaud F, Gravot A, Milesi S, Gontier E. Production of plant secondary metabolites: A historical perspective. Plant Sci 2001, 161, 839–851.

[3]   Delgoda R, Murray J. Evolutionary perspective on the role of plant secondary metabolites. In: McCreath SB, Rupika Delgoda R, ed, Pharmacognosy fundamentals applications and strategies. 1st ed San Diego, CA, USA, Academic Press, 2017, 93–100.

[4]   Salim A, Chin YW, Kinghorn A. Drug discovery from plants. In: Ramawat K, Merillon J, eds, Bioactive molecules and medicinal plants. 1st Berlin, Heidelberg, Germany, Springer, 2008, 1–24.

[5]   Cragg GM, Newman DJ. Natural Products: A continuing source of novel drug leads. Biochim Biophy Acta 2013, 1830(6), 3670–3695.

[6]   Kinghorn AD, Pan L, Fletcher JN, Chai H. The relevance of higher plants in lead compound discovery programs. J Nat Prod 2011, 24, 74(6), 1539–1555.

[7]   Veeresham C. Natural products derived from plants as a source of drugs. J Adv Pharm Technol Res 2012, 3(4), 200–201.

[8]   Dias DA, Urban S, Roessner U. A Historical overview of natural products in drug discovery. Metabolites 2012, 2, 303–336.

[9]   Pan W, Xu X, Shi N, Tsang SW, Zhang H. Antimalarial activity of plant metabolites. Int J Mol Sci 2018, 19, 1382, 10.3390/ijms19051382.

[10]   An J, Minie M, Sasaki T, Woodword JJ, Elkon KB. Anitmalarial drugs as immune modulators; New mechanisms for old drugs. Annu Rev Med 2017, 68, 317–330.

[11]   van Der Heijden R, Jacobs DI, Snoeijer W, Hallard D, Verpoorte R. The Catharanthus Alkaloids: Pharmacognosy and biotechnology. Curr Med Chem 2004, 11(5), 607–628.

[12]   Seca AML, Pinto DCGA. Plant secondary metabolites as anticancer agents: Successes in clinical trials and therapeutic application. Int J Mol Sci 2018, 19(1), 263. 10.3390/ijms19010263.

[13]   Wani MC, Taylor HL, Wall ME, Coggin P, McPhail AT. Plant antitumor agents: VI. The isolation and structure of taxol, a novel antileukemic and antitumor agent from *Taxus brevifolia*. J Am Chem Soc 1971, 93, 2325–2327.

[14]   Schiff PB, Horwitz SB. Taxol stabilizes microtubules in mouse fibroblast cells. Proc Natl Acad Sci 1980, 77(3), 1561–1565.

[15]   Wall ME, Wani MC, Cook CE, Palmer KH, McPhail AT, Sim GA. Plant antitumor agents. I. The isolation and structure of camptothecin, a novel alkaloidal leukemia and tumor inhibitor from *Camptotheca acuminata*. J Am Chem Soc 1966, 88(16), 3888–3890.

[16]   Pan L, Chai HB, Kinghorn AD. Discovery of new anticancer agents from higher plants. Front Biosci (Schol Ed) 2012, 4, 142–156.

[17]   Lichota A, Gwozdzinski K. Anticancer activity of natural compounds from plant and marine environment. Int J Mol Sci 2018, 19(11), 3533. 10.3390/ijms19113533.

[18]   Imbert TF. Discovery of podophyllotoxins. Biochimie 1998, 80(3), 207–222.

[19]   Hande KR. Etoposide: Four decades of development of a topoisomerase II inhibitor. Eur J Cancer 1998, 34(10), 1514–1521.

[20]   Kingston DG. Modern natural products drug discovery and its relevance to biodiversity conservation. J Nat Prod 2011, 74(3), 496–511.

[21]   Benatrehina PA, Pan L, Naman CB, Li J, Kinghorn AD. Usage, biological activity, and safety of selected botanical dietary supplements consumed in the United States. J Tradit Complement Med 2018, 8(2), 267–277.

[22] Tyring SK. Sinecatechins: Effects on HPV-induced enzymes involved in inflammatory mediator generation. J Clin Aesthet Dermatol 2012, 5(1), 19–26.

[23] Russo EB. Cannabis therapeutics and the future of neurology. Front Integr Neurosci 2018, 12, 51, 10.3389/fnint.2018.00051.

[24] Ming LC, Ang WC, Yang Q, Thitilertdecha P, Wong TW, Khan TM. Cosmeceuticals: Safety, efficacy and potential benefits. In: Keservani RK, Sharma AK, Kesharwani RK, ed, Recent advances in drug delivery technology. 1st Pennsylvania, USA, IGI Global, 2016, 287–308.

[25] Vaishali K, Ashwini CG, Kshitija DP, Digambar NN. Cosmeceuticals an emerging concept; A comprehensive review. Int J Res Pharm Chem 2013, 3(2), 308–316.

[26] Espinosa-Leal CA, Garcia-Lara S. Current methods for the discovery of new active ingredients from natural products for cosmeceutical applications. Planta Med 2019, 85(7), 535–551.

[27] Petruk G, Giudice R, Rigano MM, Monti DM. Antioxidants from plants protect against skin photoaging. Oxid Med Cell Longev 2018, Volume 2018, Article ID 1454936 11. pages. 10.1155/2018/1454936.

[28] Cavinato M, Waltenberger B, Baraldo G, Grade CVC, Stuppner H, Janson-Durr P. Plant extracts and natural compounds used against UV B induced photoaging. Biogerontology 2017, 18(14), 499–516.

[29] Yun K-J, Kim J-Y, Kim J-B, Lee K-W, Jeong S-Y, Park H-J, Jung H-J, Cho Y-W, Yun K, Lee K-T. Inhibition of LPS-induced NO and PGE_2 production by asiatic acid via NF-κB inactivation in RAW 264.7 macrophages: Possible involvement of the IKK and MAPK pathways. Int Immunopharmacol 2008, 8(3), 431–441.

[30] Wang C, Kao T, Lo W, Yen G. Glycyrrhizic acid and 18β-Glycyrrhetinic acid modulate lipopolysaccharide-induced inflammatory response by suppression of NF-κB through PI3K p110δ and p110γ inhibitions. J Agric Food Chem 2011, 59(14), 14, 7726–7733.

[31] Jurenka JS. Anti-inflammatory properties of curcumin, a major constituent of *Curcuma longa*: A review of preclinical and clinical research. Altern Med Rev 2009, 14(2), 141–153.

[32] Park KC, Huh SY, Choi HR, Kim DS. Biology of melanogenesis and search of hypopigmenting agents. Dermatologica Sin 2010, 28(2), 53–58.

[33] Draeloas ZD, Pugliese PT. Physiology of skin. 3rd Carol Stream, IL, USA, Alluredbooks, 2011.

[34] Hollinger JC, Angra K, Halder RM. Are natural ingredients effective in the management of hyperpigmentation? A systematic review. J Clin Aesthet Dermatol 2018, 11(2), 28–37.

[35] Pittayapruek P, Meephansan J, Prapapan O, Komine M, Ohtsuki M. Role of matrix metalloproteinases in photoaging and photocarcinogenesis. Int J Mol Sci 2016, 17(6), 868.

[36] McMullen RL. Antioxidents and the skin. Boca Raton, FL, USA, CRC Press, CRC Press 2018.

[37] Wang JJ, Sanderson BJ, Zhang W. Significant anti-invasive activities of α-mangostin from the mangosteen pericarp on two human skin cancer cell lines. Anticancer Res 2012, 32(9), 3805–3816.

[38] Losso JN, Bansode RR, Trappey A, Bawadi HA, Truax R. *In vitro* anti-proliferative activities of ellagic acid. J Nutr Biochem 2004, 15(11), 672–678.

[39] Pacheco-Palencia LA, Noratto G, Hingorani. L, Talcott ST, Mertens-Talcott SU. Protective effects of standardized pomegranate (*Punica granatum* L.) polyphenolic extract in ultraviolet-irradiated human skin fibroblasts. J Agric Food Chem 2008, 56(18), 8434–8441.

[40] Park JS, Kim DK, Shin H-D, Lee HJ, Jo HS, Jeong JH, Choi YL, Lee CJ, Hwang S. Apigenin regulates interleukin-1β-induced production of matrix metalloproteinase both in the knee joint of rat and in primary cultured articular chondrocytes. Biomol Ther (Seoul) 2016, 24(2), 163–170.

[41] Shankar E, Goel A, Gupta K, Gupta S. Plant flavone apigenin: An emerging anticancer agent. Curr Pharmacol Rep 2017, 3(6), 423–446.

[42] Imokawa G, Ishida K. Biological mechanisms underlying the ultraviolet radiation-induced formation of skin wrinkling and sagging I: Reduced skin elasticity, highly associated with enhanced dermal elastase activity, triggers wrinkling and sagging I. J Mol Sci 2015, 16, 7753–7775.

[43] Thring TSA, Hili P, Naughton DP. Anti-collagenase, anti-elastase and anti-oxidant activities of extracts from 21 plants. BMC Complement Altern Med 2009, 9, 27, 10.1186/1472-6882-9-27.

[44] Safayhi H, Rall B, Sailer ER, Ammon HP. Inhibition by boswellic acids of human leukocyte elastase. J Pharmacol Exp Ther 1997, 281(1), 460–463.

[45] Ammon HP, Mack T, Singh GB, Safayhi H. Inhibition of leukotriene B4 formation in rat peritoneal neutrophils by an ethanolic extract of the gum resin exudate of *Boswellia serrata*. Planta Med 1991, 57(3), 203–207.

[46] Tausch L, Henkel A, Siemoneit U, Poeckel D, Kather N, Franke L, Hofmann B, Schneider G, Angioni C, Geisslinger G, Skarke C, Holtmeier W, Beckhaus T, Karas M, Jauch J, Werz O. Identification of human cathepsin G as a functional target of boswellic acids from the anti-inflammatory remedy frankincense. J Immunol 2009, 183(5), 3433–3442.

[47] Reddy GK, Chandrakasan G, Dhar SC. Studies on the metabolism of glycosaminoglycans under the influence of new herbal anti-inflammatory agents. Biochem Pharmacol 1989, 38(20), 3527–3534.

[48] Karabın M, Hudcová T, Jelinek L, Dostálek P. Biologically active compounds from hops and prospects of their use. Compr Rev Food Sci Food Saf 2016, 15, 542–567.

[49] Verzele M. 100 years of hop chemistry and its relevance to brewing. J Inst Brew 1986, 92, 32–48.

[50] Negrão R, Costa R, Duarte D, Gomes TT, Mendanha M, Moura L, Vasques L, Azevedo I, Soares R. Angiogenesis and inflammation signaling are targets of beer polyphenols on vascular cells. J Cell Biochem 2010, 111(5), 1270–1279.

[51] Domingo DS, Camouse MM, Hsia AH, Matsui M, Maes D, Ward NL, Cooper KD, Baron ED. Anti-angiogenic effects of epigallocatechin-3-gallate in human skin. Int J Clin Exp Pathol 2010, 3(7), 705–709.

[52] Amaro-Ortiz A, Yan B, D'Orazio JA. Ultraviolet Radiation, aging and the skin: Prevention of damage by topical cAMP manipulation. Molecules 2014, 19(5), 6202–6219.

[53] Passeron T, Namiki T, Passeron HJ, Pape EL, Hearing VJ. Forskolin protects keratinocytes from UVB-induced apoptosis and increases DNA repair independent of its effects on melanogenesis. J Invest Dermatol 2009, 129(1), 162–166.

Part II: **Introduction to plant secondary metabolites**

Chandani Ranasinghe
3 Plant phenolic compounds

3.1 Introduction

Phenolic compounds are the most abundant and widely distributed secondary metabolites in plants. They have a vast chemical diversity [1]. The occurrence of these compounds is reported in bacteria, fungi, and algae also [2–5]. Higher plants produce several thousands of phenolic compounds and the number of those which are being identified and characterized increases continuously.

Phenolic compounds are characterized by having at least one hydroxyl (OH) group attached to a benzene ring or a complex aromatic structure. Therefore, the parent structure of this group of substances is phenol (**1**). Structurally they vary from a single ring phenol to polyphenols where several phenolic units are present. Based on the number of OH groups, they can be classified as mono-, di-, tri-, and polyhydric phenols. Depending on the number of phenolic units present, they can be classified as monomeric, dimeric, or polymeric phenols. Phenolic compounds show huge chemical diversity due to the other functional derivatives such as esters. ethers, amides, and glycosides present in them. Among the plant phenolic compounds known, flavonoids form the largest group. Quinones, xanthones, and coumarins are other monomeric plant phenolics, while lignans are dimeric phenolics. Melanins, tannins, and lignins are examples of polymeric phenolics [1].

Phenolic compounds play a variety of roles in plants. Many of them are structural constituents in cell walls that provide mechanical support to plants. Other nonstructural phenolic constituents play important roles in plant growth and survival. The majority of phenolics are related to defense responses against herbivores and pathogens. Some help in accelerating pollination, camouflage, and regulating the growth of nearby competing plants [6].

The term "plant phenolics" strictly refers to the secondary metabolites which originate from the shikimic acid pathway or polyketide pathway. According to this definition, phenolics that are of terpenoid origin (e. g. carvacrol) are not included in this group [7].

Based on the basic skeleton, phenolic compounds can be categorized into several groups as shown in Table 3.1 [1].

Chandani Ranasinghe, Department of Chemistry,The Open University of Sri Lanka, Nawala, Nugegoda, Sri Lanka, e-mail: crana@ou.ac.lk

https://doi.org/10.1515/9783110595949-003

Table 3.1: Structural types of phenolic compounds.

No. of C atoms	Basic skeleton	Class	Examples
6	C_6	Simple phenols, benzoquinones	Catechol (**2**), hydroquinone, 2,6-dimethoxybenzoquinone
7	C_6-C_1	Phenolic acids, phenolic aldehydes	Gallic acid (**7**), salicylic acid (**9**)
8	C_6-C_2	Acetophenones, tyrosine derivatives, phenylacetic acid	3-Acetyl-6-methoxybenzaldehyde, tyrosol (**67**), p-hydroxyphenylacetic acid, homogentisic acid
9	C_6-C_3	Hydroxycinnamic acid, phenylpropenes, coumarins, isocoumarins, chromenes	Caffeic acid (**14**), ferulic acid (**15**), myristicin, eugenol (**19**), umbelliferone (**21**), aesculetin (**22**), bergenin (**23**), eugenin (**24**)
10	C_6-C_4	Naphthoquinones	Juglone (**25**), plumbagin
n > 12	$(C_6-C_3)_n$ $(C_6)_n$ $(C_6-C_3-C_6)_n$	Lignins Melanins Condensed tannins (flavolans)	Raspberry ellagitannin, Tannic acid
13	$C_6-C_1-C_6$	Xanthonoids	Mangiferin
14	$C_6-C_2-C_6$	Stilbenoids, anthraquinones	Resveratrol (**29**), emodin (**32**)
15	$C_6-C_3-C_6$	Chalconoids, flavonoids, isoflavonoids, neoflavonoids	Quercetin (**34**), cyanidin, genistein (**44**)
18	$(C_6-C_3)_2$	Lignans, neolignans	Pinoresinol, eusiderin
30	$(C_6-C_3-C_6)_2$	Bioflavonoids	Amentoflavone

(a) C_6 phenolic compounds: simple phenols

Although these simple phenols such as phenol (**1**), catechol (**2**), and pyrogallol (**3**) are rare to find in nature, their residues occur in more complex molecules like flavonoids.

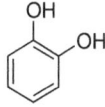

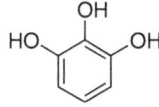

Phenol (**1**)   Catechol (**2**)   Pyrogallol (**3**)

(b) C_6–C_1 phenolic compounds: phenolic acids

Phenolic acids such as *p*-hydroxybenzoic acid (**4**), protocatechuic acid (**5**), vanillic acid (**6**), gallic acid (**7**), syringic acid (**8**), salicylic acid (**9**), and gentisic acid (**10**) are widely distributed in higher plants. Some of these phenolic acids are found as constituents in lignin and tannin.

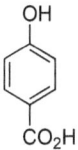

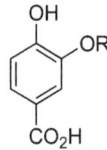

p-Hydroxybenzoic acid (**4**)

R= H: Protocatechuic acid (**5**)

R= CH_3: Vanillic acid (**6**)

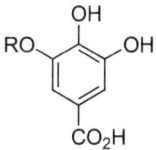

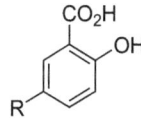

R= H: Gallic acid (**7**)

R=CH_3: Syringic acid (**8**)

R= H: Salicylic acid (**9**)

R= OH: Gentisic acid (**10**)

(c) C_6–C_2 phenolic compounds: acetophenones and phenylacetic acids

Phenolic compounds belonging to this group are not very common. They can be found as acetophenones and phenylacetic acids and derivatives of them.

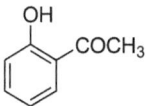

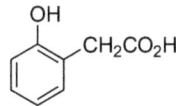

2-Hydroxyacetophenone (**11**)

2-Hydroxyphenylacetic acid (**12**)

(d) C_6–C_3 phenolic compounds: hydroxycinnamic acids, phenylpropenes, coumarins, isocoumarins, and chromones

C_6–C_3 is the most abundant and most important skeleton of phenolic compounds. These can be subdivided into three groups.

i. Hydroxycinnamic acids and related compounds

Some examples of widely distributed cinnamic acid derivatives are p-coumaric acid (**13**), caffeic acid (**14**), ferulic acid (**15**) and sinapic acid (**16**). Alcohols such as coniferyl alcohol (**17**) and sinapyl alcohol (**18**), derived from cinnamic acid are constituents of woody plants. They are precursors of lignin.

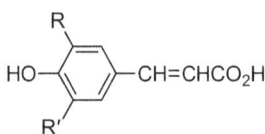

R= R'= H: p-Coumaric acid (**13**)

R= OH, R'= H: Caffeic acid (**14**)

R= OCH$_3$, R'= H: Ferulic acid (**15**)

R= R'= OCH$_3$: Sinapic acid (**16**)

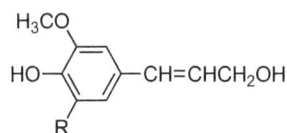

R= H: Coniferyl alcohol (**17**)

R= OCH$_3$: Sinapyl alcohol (**18**)

ii. Phenylpropenes

Eugenol (**19**) and isoeugenol (**20**) are two examples of phenylpropenes. They are constituents of essential oils.

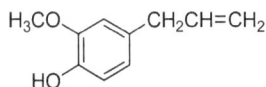

Eugenol (**19**)

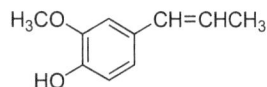

Isoeugenol (**20**)

iii. Coumarins, isocoumarins, and chromones

In this group of phenolics, C_3 chain of the C_6–C_3 skeleton is present in the form of an oxygen heterocycle. Given below are some examples.

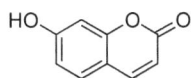

Umbelliferone (**21**)
(coumarin)

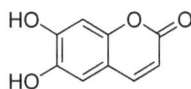

Aesculetin (**22**)
(coumarin)

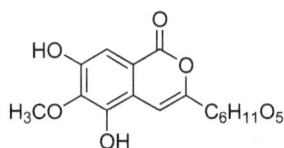

Bergenin (**23**)
(isocoumarin)

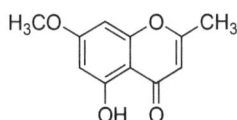

Eugenin (**24**)
(chromene)

(e) C₆–C₄ phenolic compounds: naphthoquinones

Naphthoquinones have the basic skeleton of naphthalene.

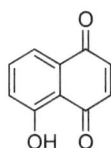

Juglone (**25**)

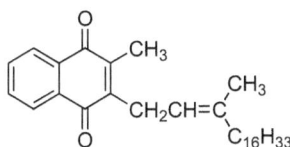

Vitamin K (**26**)

(f) C₆–C₁–C₆ phenolic compounds: benzophenones

Benzophenones and xanthones have this skeleton where two phenyl groups are linked through a carbonyl function.

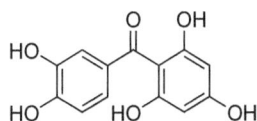

Maclurin (**27**)
(benzophenone)

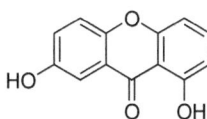

Euxanthone (**28**)
(xanthone)

(g) C₆–C₂–C₆ phenolic compounds: stilbenes and anthraquinones

This subgroup includes stilbenes that have a central ethylene moiety with two phenyl groups at each end and anthraquinones which are derivatives of anthracene.

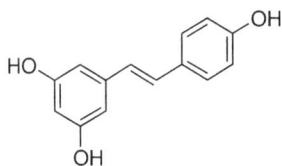

Resveratrol (**29**)
(stilbene)

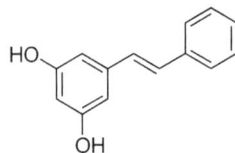

Pinosylvin (**30**)
(stilbene)

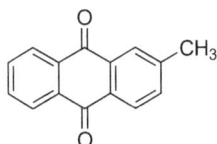

Tectoquinone (**31**)
(anthraquinone)

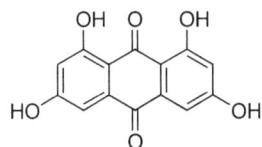

Emodin (**32**)
(anthraquinone)

(h) C_6–C_3–C_6 phenolic compounds: flavonoids and isoflavonoids

i. Flavonoids

The flavonoid group consists of several different subgroups, namely, flavones, flavonols, flavonones, flavononols, chalcones, dihydrochalcones, aurones, catechins, anthocyanidins, and leucoanthocyanidins. This grouping is based on the nature of the C_3 moiety of the molecule. C_6 is always a benzene ring.

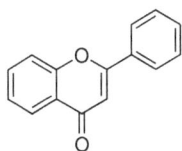

Flavone (**33**)

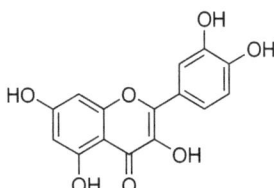

Quercetin (**34**)
(flavonol)

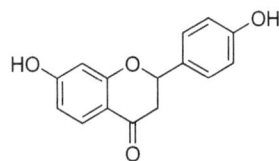

Naringenin (**35**)
(flavanone)

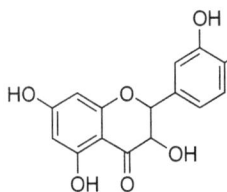

Taxifolin (**36**)
(flavanonol)

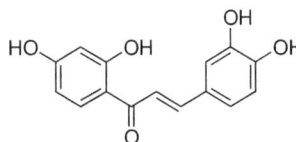

Butein (**37**)
(chalcone)

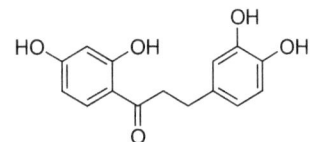

Phloretin (**38**)
(dihydrochalcone)

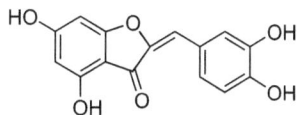

Aureusidin (**39**)
(aurone)

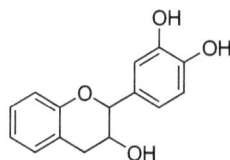

Flavan-3-ol (**40**)
(catechin)

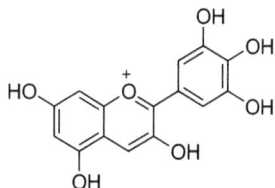

Delphinidin (**41**)
(anthocyanidin)

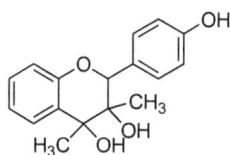

Leucopelargonidin (**42**)
(Leucocyanidin or flavan-3,4-diol)

ii. Isoflavonoids

The distribution of isoflavonoids is lesser than flavonoids. These are isomers of flavone
(**33**) where ring B is attached to the carbon atom at position 3, instead of position 2.

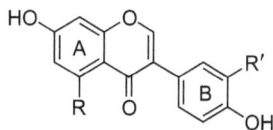

R= R′= H: Daidzein (**43**)

R= OH, R′= H: Genistein (**44**)

R= R′= OH: Orobol (**45**)

(i) (C_6–C_3)$_2$ phenolic compounds: lignans

Lignans are dimeric phenylpropanoids but their chemical structures are diverse
and complex. C_6–C_3 units are linked by central carbons of their side chains. They
are of different structural types.

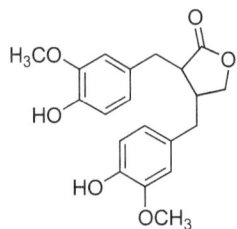

Matairesinol (**46**)   Podophyllotoxin (**47**)

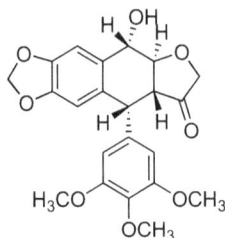

(j) (C₆–C₁–C₆)₂ phenolic compounds: biflavonoids

These are dimers of identical or non-identical flavonoid units. They are joined in a symmetrical or unsymmetrical manner through an alkyl or an alkoxy linker.

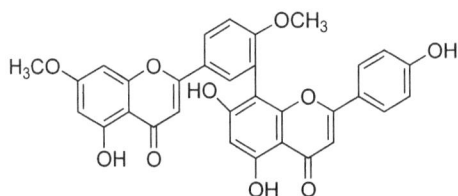

Ginkgetin (**48**)

(k) (C₆)ₙ phenolic compounds: melanins

Melanins mostly have relatively diverse and undefined structures. In plants, the most common precursor of melanin is catechol (**2**). Therefore, the melanin that is formed by oxidative polymerization is called catechol melanin.

(l) (C₆–C₃)ₙ phenolic compounds: lignins

Lignins are complex polymers of phenylpropanoid units. The precursors of lignins are identified as various substituted cinnamyl alcohols such as *p*-coumaryl alcohol, coniferyl alcohol (**17**) and sinapyl alcohol (**18**).

(m) (C₆–C₃–C₆)ₙ phenolic compounds: tannins

Tannins are water-soluble high molecular weight polyphenolic compounds that can bind with other macromolecules such as proteins, sugars, and cellulose. Tannins in higher plants can be divided into two subgroups as hydrolysable tannins and condensed tannins.

Hydrolysable tannins consist of a central core of glucose or other polyhydric alcohol esterified with gallic acid (**7**); gallotannins (**49**) or ellagic acid; ellagitannins. They are hydrolysed by weak acids or bases.

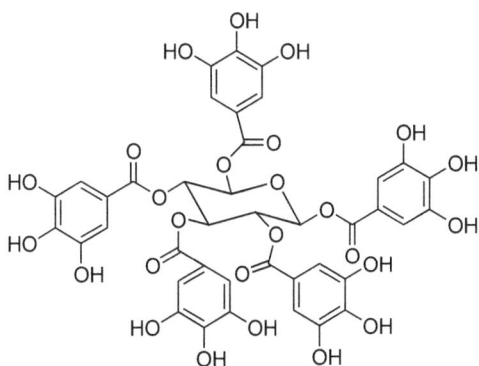

Gallotannin (**49**)

Condensed tannins are more distributed in the plant kingdom than hydrolysable tannins. They are also called proanthocyanidins as they yield anthocyanidin on depolymerization under oxidative conditions. Condensed tannins are polymers (2 to 60 monomeric units) of flavonoids, linked through C–C bonds which cannot be cleaved by hydrolysis. They have complex structures due to structurally different flavonoid units forming links through various sites.

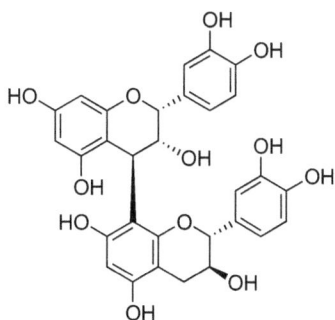

Procyanidin B2 (**50**)

3.2 Biosynthesis of phenolic compounds

Phenolic compounds are secondary metabolites. They are synthesized in plants from simple, low molecular weight primary metabolites (e.g., simple sugars, simple organic acids, and amino acids). Phenolic compounds in plants are derived via three routes. They are,

- Polyketide pathway
- Shikimic acid pathway
- Mixed biosynthetic pathway

The shikimic acid pathway starts from carbohydrates and produces phenolic compounds with hydroxyl substituents at 3, 4 positions (catechol type) or 3, 4, 5 positions (pyrogallol type) of the benzene ring. Polyketide pathway arises from acetyl coenzyme A and malonyl coenzyme A and gives phenolics with hydroxyl substituents at 1,3 positions (resorcinol type) or 1, 3, 5 positions (phloroglucinol type). Therefore, by looking at the hydroxylation pattern in the aromatic ring of the phenolic compound, one can determine the biosynthetic pathway it had originated from.

Some phenolics with more than one phenolic nucleus (e.g. flavonoids) are resultant from the mixed biosynthetic pathway. They contain aromatic rings derived from both the polyketide and shikimic acid pathways. Pathways by which phenolic compounds are synthesized in plants are summarized in Figure 3.1 [8–10].

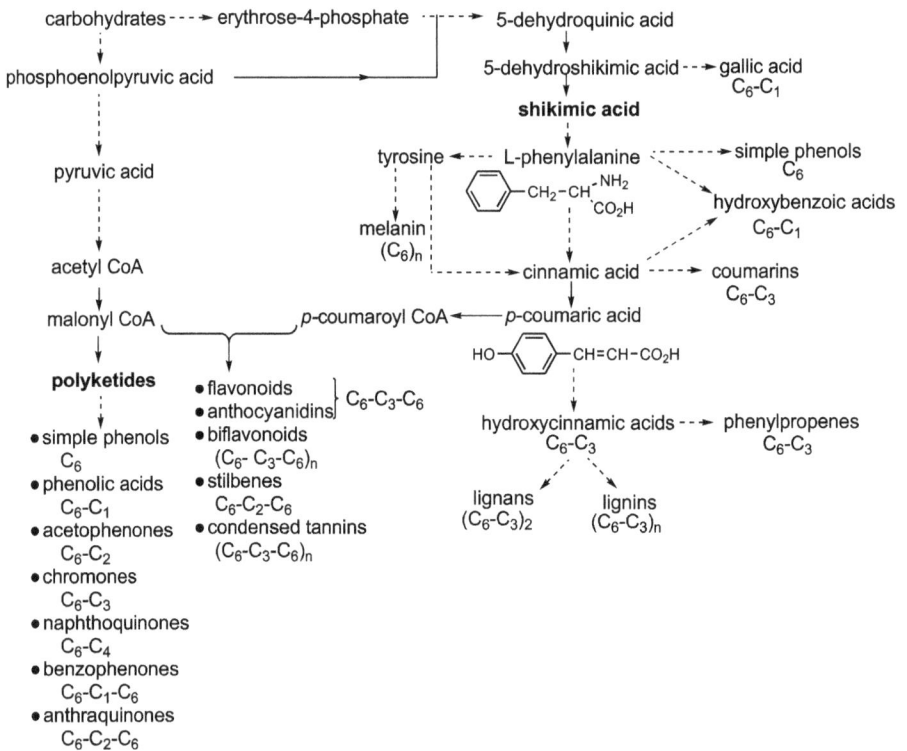

Figure 3.1: Biosynthetic pathways of plant phenolic compounds.

3.2.1 Biosynthesis of phenolic compounds via polyketide pathway

The formation of phenolic compounds by the polyketide pathway starts from the condensation of acetyl and malonyl coenzyme A (Figure 3.2). Malonyl coenzyme A is formed by carboxylation of acetyl co-enzyme A. The $-CH_2-$ group in malonyl coenzyme A is activated due to the presence of carbonyl groups on either side. It easily undergoes Claisen type condensation with acetyl coenzyme A to give acetoacetyl coenzyme A. Further condensation with malonyl units gives polyketide chains (poly β-ketoesters) of various lengths [8–10].

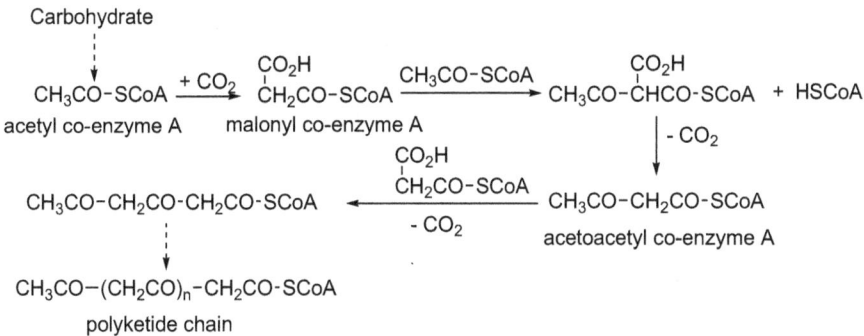

Figure 3.2: Pathway for the formation of polyketide chain.

Polyketide chain can fold and undergo cyclization (intramolecular aldol condensation) to produce an aromatic nucleus with meta-oriented (alternate) hydroxyl groups. The biosynthesis of phloroglucinol given in Figure 3.3 illustrates this fact.

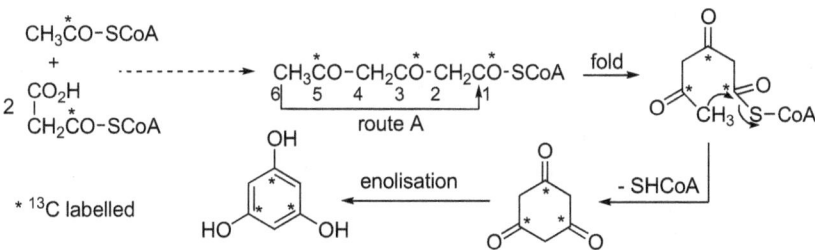

Figure 3.3: Biosynthesis of phloroglucinol.

Cyclization of the polyketide chain can take place in several ways. Four different cyclization pathways of tetraketide are given below (Figure 3.4).

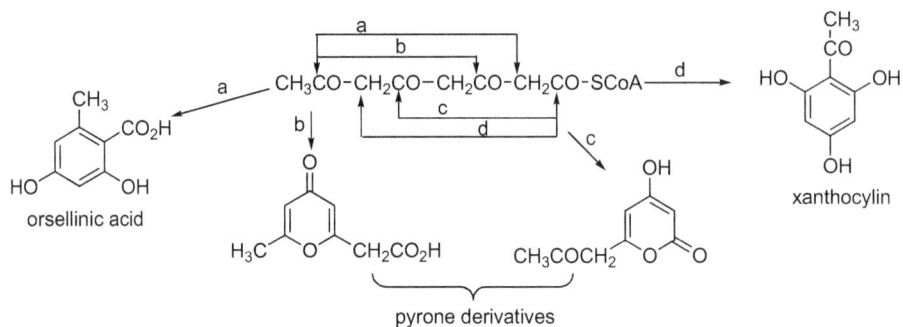

Figure 3.4: Different cyclization pathways of tetraketide chain.

Biosynthesis of quinones and chromones show different ways of folding the poly-ketide chain. Many other reactions such as oxygenation, deoxygenation, hydration, dehydration, decarboxylation, O-methylation, etc. also take place during the cycliza-tion process.

3.2.2 Biosynthesis of phenolic compounds via shikimic acid pathway

The precursors for the shikimate pathway to produce aromatic compounds are phos-phoenolpyruvic acid and erythrose-4-phosphate which are resultant from glycolysis (Figure 3.5). Shikimic acid is one of the key intermediates in this pathway [8–10].

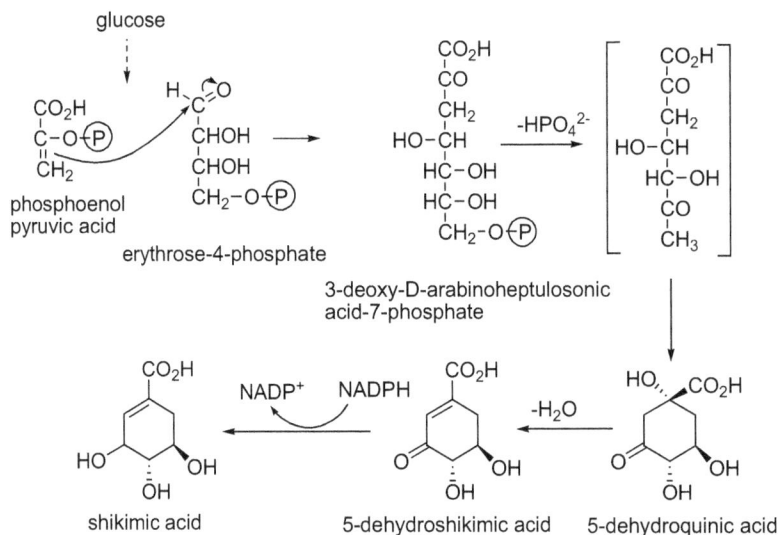

Figure 3.5: Formation of shikimic acid from glucose.

The major end-product of the shikimate pathway is phenylalanine which is pro-duced from the intermediate, shikimic acid. This conversion involves the aromatiza-tion of the cyclohexane ring in three easy steps. Phenylalanine is the starting material for many important organic compounds including phenolics. Several C_6–C_3 phenolic compounds such as p-coumaric acid (**13**), caffeic acid (**14**), ferulic acid (**15**), and si-napic acid (**16**) and their corresponding alcohols are formed by deamination and en-zymatic oxidation of phenylalanine (Figure 3.6).

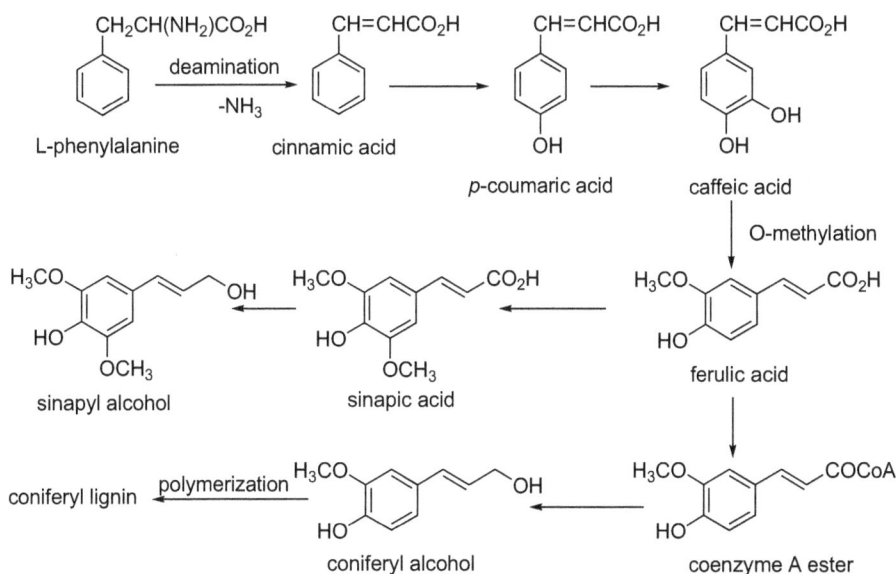

Figure 3.6: Formation of phenolic compounds from phenylalanine.

Coumarins are formed by o-hydroxycinnamic acid derivatives by lactonization (cyclic ester formation). Oxidative dimerization and polymerization of phenylpropa-noid free radicals (C_6–C_3) produce lignans and lignins.

3.2.3 Biosynthesis of flavonoids via the combined pathway

Flavonoids are C_6–C_3–C_6 phenolic compounds where the two benzene rings are derived from two different biosynthetic pathways: 1,3- or 1,3,5-hydroxylated (phloroglucinol type) ring arises from polyketide pathway while 4-, 3,4- or 3,4,5-hydroxylated ring arises from shikimic acid pathway [8–10].

Joining of a C_9 fragment derived from shikimic acid and C_6 fragment derived from polyketide chain to form flavonoids can be given as follows (Figure 3.7). Chal-cone is a common intermediate in the synthesis of derivatives of flavonoid.

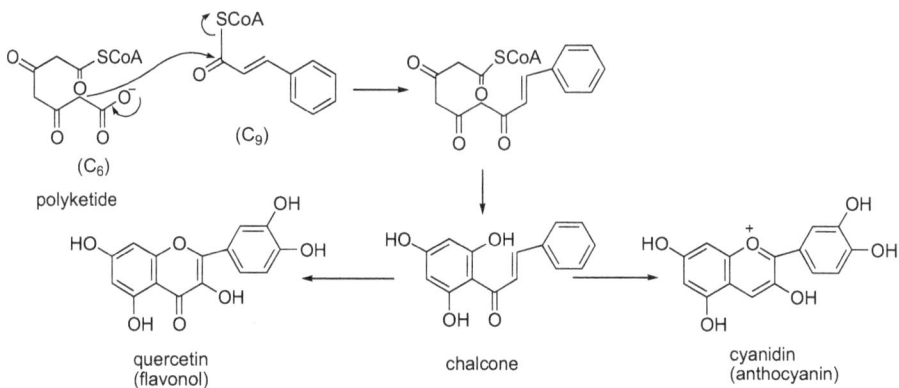

Figure 3.7: Joining of C_9 and C_3 fragments to form chalcone and derivatives.

3.3 Pharmacological importance of phenolic compounds

Phenolic compounds are found to exhibit a number of pharmacological properties. The best known is the simple phenolic compound salicylic acid (**9**) which is used as a drug in its acetate form (acetylsalicylic acid) which is widely used as an analgesic in the treatment of pain, fever, and inflammation. This phenolic compound is found in its glycoside form, β-D-salicin in various plants including willow tree (*Salix alba*) from which it was first discovered. In the human gastrointestinal tract, β-D-salicin is biotransformed to salicylic acid (**9**) (Figure 3.8) [11].

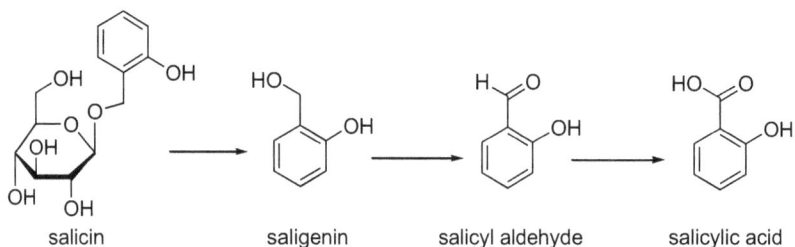

Figure 3.8: Biotransformation of salicin to salicylic acid.

In phenolic compounds, the characteristic functional group is the hydroxyl group attached to benzene. Biological and pharmacological activities shown by phenolic compounds are not always attributable to the phenol OH group. Different biological and pharmacological activities are rendered by other functionalities attached to the phenolic compounds except for antioxidant properties exhibited by them.

Phenolic compounds can act as antioxidants due to the hydrogen donor ability of the phenyl bearing hydroxyl group. They can react with the free radicals (reactive oxygen and reactive nitrogen species produced from either endogenous or exogenous sources) which can oxidatively damage lipids, proteins, and nucleic acids. Free radical thus formed is extra stabilized due to delocalization in the π-electron system of the benzene ring.

This antioxidant ability in the form of radical scavenging or metal ion chelation is beneficial to human health as it can reduce oxidative damage to the cell constituents. Oxidative damage will subsequently lead to diseases like cancer, liver disease, premature aging, inflammation, diabetes, arthritis, Alzheimer's disease, Parkinson's disease, and atherosclerosis [12–14].

3.3.1 Cardioprotective effect

Cardiovascular diseases include several conditions that affect the structure and functions of the heart or blood vessels. Atherosclerosis (building up of plaque inside arteries) is the main underlying cause of cardiovascular diseases.

Studies have shown that oxidation of LDL which is the main reason for developing atherosclerosis is inhibited by polyphenols. The antioxidant, anti-platelet, and anti-inflammatory effects of polyphenols, as well as the ability to increase HDL levels, improve functions of the endothelium [15].

The flavonoid quercetin (**34**) and tea catechins are reported to be contributing to reducing atherosclerosis through various mechanisms. It is also shown that resveratrol (**29**), a wine polyphenol, reduces the chances of cardiovascular diseases. In general, recent research has shown that a polyphenolic-rich diet reduces the risk of myocardial infarction [15].

3.3.2 Anti-cancer effect

"Cancer" is the term given to a collection of diseases. In all these types uncontrolled cell division takes place which eventually spreads into surrounding tissues.

The pharmacological importance of phenolic compounds is explored in several anti-cancer studies. Extracts of edible berries that are rich in polyphenols have been tested on various cancer cell lines for anti-cancer activity and have shown their effectiveness in different stages of cancer. *In vitro* studies on phenolic extracts containing anthocyanins, kaempferol (**55**), quercetin (**34**), esters of *p*-coumaric acid (**13**), and ellagic acid have inhibited the growth of human oral, breast, colon, and prostate tumor cell lines. Ellagitannins have shown inhibition of cancer cell growth while procyanidins have also exhibited antiproliferative activity. It is postulated that many phenolic compounds are capable of inactivating NF-kB-dependent signaling which is responsible for the activation of many genes involved in cell proliferation [16, 17].

Some isolated polyphenols [quercetin (**34**), resveratrol (**29**), catechin, and (-)-epicatechin (**51**)], tea extract and major green tea polyphenols [(-)-epicatechin (**51**), (-)-epigallocatechin (**63**), (-)-epicatechin-3-gallate, and (-)-epigallocatechin-gallate] have exhibited anti-cancer activity at various concentration levels depending on the system and the test substance. Particularly, resveratrol (**29**) was capable of suppressing angiogenesis and metastasis while multiple pathways involved in cell growth, apoptosis, and inflammation were modulated by this compound [15]. A number of flavones [(apigenin (**54**), baicalein, luteolin (**56**) and rutin (**58**)], flavonones [hesperidin (**59**) and naringin (**60**)], and sesame lignans [sesamin (**61**), sesaminol (**62**) and episesamin] have shown anti-cancer effects against different cancer cell lines including colon, prostate, leukemia, liver, cervix, pancreas, and breast [17].

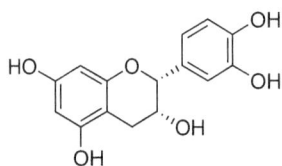

(-)-Epicatechin (**51**)

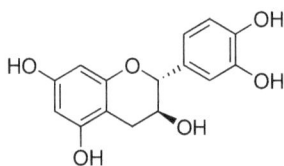

(+)-Catechin (**52**)

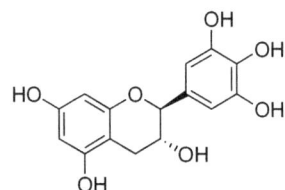

(-)-Gallocatechin (**53**)

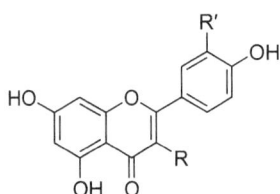

R=R'= H: Apigenin (**54**)

R=OH, R'= H: Kaempferol (**55**)

R=H, R'= OH: Luteolin (**56**)

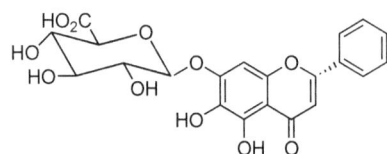

Baicalin (**57**)

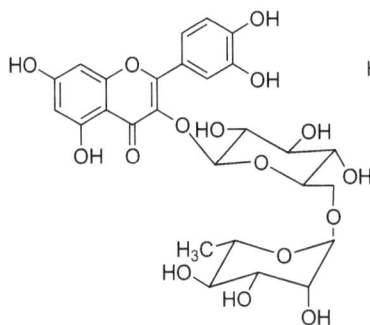

Rutin (**58**)

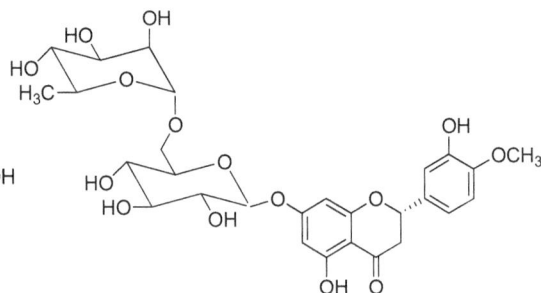

Hesperidin (**59**)

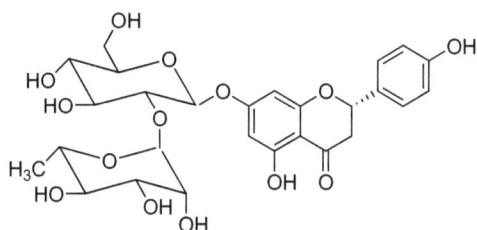

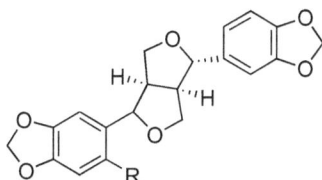

Naringin (**60**)

R=H: Sesamin (**61**)

R=OH: Sesaminol (**62**)

3.3.3 Anti-diabetic effect

Diabetes is a condition where the blood glucose level is too high which can be due to two reasons: inability to produce insulin by the body (Type 1) or ineffectiveness of insulin in the body (Type 2). Insulin hormone is the chemical messenger that allows cells to absorb glucose which regulates the glucose level in the blood.

Flavonoids, phenolic acids, and tannins have shown anti-diabetic activity by way of inhibiting α-amylase and α-glucosidase enzymes which are responsible for the digesting of dietary carbohydrates into glucose [18]. This helps to decrease the postprandial hyperglycemia by impeding the absorption of glucose and would be an effective strategy for treating type 2 diabetes [19].

Investigation of plants used in traditional and Ayurvedic medicine has revealed that (-)-catechin gallate (**65**) and (-)-epicatechin gallate (**66**) are responsible for inhibition of α-amylase and α-glucosidase enzymes. Hydroxyl groups present in phenolic compounds can form H-bonds with polar groups in enzymes. Hydrophobic sites are present in enzymes and hydrophobic galloyl groups are found in polyphenols. The binding of enzymes can take place through these hydrophobic associations and the interaction between galloyl moiety, and the enzymes affect the effectiveness of these digestive enzymes [19].

Various studies have shown anti-diabetic action of (-)-epicatechin (**51**), (+)-catechin (**52**), (-)-epigallocatechin (**63**), (-)-epicatechin gallate (**66**), isoflavones from soybeans, tannic acid, glycyrrhizin from licorice root, chlorogenic acid, saponins, stilbenes, and resveratrol (**29**) (trans-3,5,4′-trihydroxystilbene) through various mechanisms. Onion polyphenols, especially quercetin (**34**), *Hibiscus sabdariffa* extract which contains polyphenolic acids, flavonoids, protocatechuic acid (**5**), and anthocyanins are also found to possess strong antidiabetic activity [15].

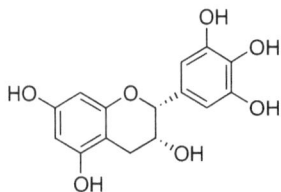

(-)-Epigallocatechin (**63**)

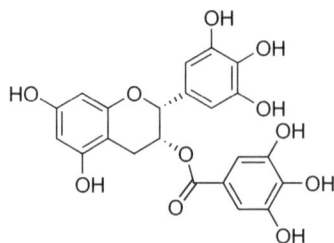

(-)-Epigallocatechin gallate (**64**)

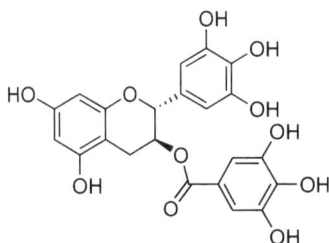

(-)-Catechin gallate (**65**)

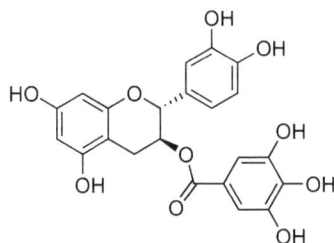

(-)-Epicatechin gallate (**66**)

The fruit of *Punica granatum* (pomegranate) is rich in flavonoids (flavonols and flavanols), anthocyanins, hydrolysable tannins [ellagitannins, gallotannins (**49**)], condensed tannins (proanthocyanidins), and organic phenolic acids have shown to be effective in type-2 diabetes. Its mechanism of action is related to decreasing lipid peroxidase and oxidative stress through various mechanisms, for example, by enhancing the antioxidant activity of some enzymes, inducing metal chelating activity, or either inhibiting or activating transcriptional factors [18].

3.3.4 Anti-neurodegenerative effect

Diseases caused by the progressive death of the neurons in different regions of the nervous system are known as neurodegenerative diseases where Alzheimer's disease and Parkinson's disease are the most common types.

Polyphenols have shown promising activity against neurodegenerative diseases. The main cause of these diseases is oxidative stress and damage to brain macromolecules. Polyphenols act as antioxidants that prevent oxidation of proteins, lipid peroxidation, and generation of reactive oxygen (ROS) species. They also act as anti-inflammatory and anti-apoptotic (active against programmed cell death) agents. Contribution by resveratrol (**29**) as a neuroprotective metabolite is widely studied; however, the detailed mechanisms underlying this effect have not been investigated yet [20].

The neuroprotective ability of water-soluble polyphenols in wine (phenolic acids, stilbenes, tannins, flavonoids, flavanols, and anthocyanins) has been studied, and it was found that the mechanism of action involves their antioxidant activity through scavenging intracellular ROS and inhibition of LDL oxidation [21].

Resveratrol (**29**), which shows antioxidant activity by scavenging oxygen and lipid hydroperoxyl free radicals, is reported to decrease the incidence of Alzheimer's disease. Besides, some findings revealed that resveratrol (**29**) is beneficial in sustaining healthy nerves and important brain functions including cognitive processes [22]. Consumption of fruit and vegetable juices of the high content of polyphenols is also found to delay the onset of this disease. Maize bran polyphenol ferulic acid (**15**) is also found to be effective in reducing the risk of Alzheimer's disease, and this is attributed to the antioxidant and anti-inflammatory properties of the compound [15].

The therapeutic role of green tea and catechins in Parkinson's disease can be attributed to their ability to chelate iron, a property that contributes to the antioxidant activity by preventing redox-active transition metal from catalyzing the formation of free radicals [15].

3.3.5 Anti-aging effect

Biologically, aging is defined as the deterioration of physiological functions necessary for survival and fertility. It is the impact of the accumulation of a wide variety of molecular and cellular damage to the body over time. This gradually leads to a decrease in physical and mental capacity and an increased risk of diseases that ultimately bring about death [23].

Several studies suggested that the combination of antioxidant/anti-inflammatory polyphenols in fruits and vegetables could be effective as anti-aging compounds. For example, anthocyanins were found to be potent antioxidant/anti-inflammatory agents as well as inhibitors of lipid peroxidation and inflammatory mediators of cyclooxygenase 1 and 2 (COX-1 and COX-2; two enzymes producing prostaglandins that promote inflammation, pain, and fever). Quercetin (**34**) and the grape polyphenol resveratrol (**29**) have also shown promising anti-aging effects [15].

During aging, the extracellular matrix proteins like collagen and elastin become susceptible to the excessive activity of proteolytic enzymes-matrix metalloproteinases (MMPs) collagenase, and elastase. Under normal physiological conditions, the activity of these enzymes is regulated; however, the imbalance in homeostasis leads to loss of integrity of the skin tissue, resulting in the formation of wrinkles. Many polyphenols from plants (e.g., cocoa) functioned as inhibitors of collagenases and elastases, thus would be beneficial in the maintenance of proper skin structure. Moreover, polyphenols from green tea, (catechin, epigallocatechin gallate) were formulated into anti-aging skin-care products with restrained collagenase and elastase inhibition [24].

3.3.6 Effect on other human diseases

Many studies report the beneficial effects of polyphenolic extracts or diet on various medical issues. Dietary phenolic acids are inversely associated with hypertension irrespective of their dietary source [25]. In a population-based study in Brazil, it was found that the polyethanoid tyrosol (**67**) and some classes of polyphenols such as alkylphenols, lignans, and stilbenes exert a beneficial effect in treating hypertension [26].

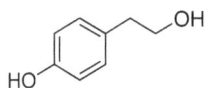

Tyrosol (**67**)

Silymarin, which is a mixture of flavonoids and polyphenols, exerts membrane-stabilizing and antioxidant activity as well as promotes hepatocyte regeneration, reduces the inflammatory reaction, and inhibits the fibrogenesis in the liver. Thus, it has the potential for preventing and curing liver diseases [27].

Some selected polyphenols [green tea polyphenols, grape seed proanthocyanidins, resveratrol (**29**), silymarin, and genistein (**44**)] have been reported to protect skin from adverse effects of UV radiation, including the risk of skin cancers. Polyphenols may supplement sunscreen protection and may be useful for skin diseases associated with solar radiation–induced inflammation, oxidative stress, and DNA damaging. Thus, the topical application of polyphenols would be beneficial due to their photoprotective effects [28].

Polyphenolic extracts of *Euphorbia umbellata* bark used in Brazilian folk medicine for gastric problems had shown anti-ulcer effects. Polyphenols present in the extract were capable of acting on the cyclooxygenase pathway and thereby promoting the maintenance of prostaglandin levels thus, exerting gastroprotection effect. Similarly, by increasing the nitric oxide levels these compounds could contribute to the protection of gastric mucosa while their involvement of the protein components of the glutathione complex was also related to the potential anti-ulcer action [29].

3.4 Conclusion

Multiple pharmacological functions exerted by phenolic compounds signify their suitability as prospective lead compounds for novel pharmaceuticals.

References

[1]   Harborne JB. Plant phenolics. Bell EA, Charlwood BV, eds, Encyclopedia of Plant Physiology, Volume 8- Secondary Plant Products. Berlin/ Heidelberg,Germany, Springer -Verlag, 1980, 329–395.

[2]   Babu B, Wu JT. Production of natural butylated hydroxytoluene as an antioxidant by freshwater phytoplankton. J Phycol 2008, 44(6), 1447–1454.

[3]   Barros LI, Dueñas M, Ferreira IC, Baptista P, Santos-Buelga C. Phenolic acids determination by HPLC–DAD–ESI/MS in sixteen different Portuguese wild mushrooms species. Food Chem Toxicol 2009, 47(6), 1076–1079.

[4]   Del Signore A, Romeo F, Giaccio M. Content of phenolic substances in basidiomycetes. Mycol Res 1997, 101(5), 552–556.

[5]   Onofrejová L, Vasíčková J, Klejdus B, Stratil P, Misurcová L, Krácmar S, Kopecký J, Vacek J. Bioactive phenols in algae: The application of pressurized-liquid and solid-phase extraction techniques. J Pharm Biomed Anal 2010, 51(2), 464–470.

[6]   Lin D, Xiao M, Zhao J, Li Z, Xing B, Li X, Kong M, Li L, Zhang Q, Liu Y, Chen H, Qin W, Wu H, Chen S. An Overview of plant phenolic compounds and their importance in human nutrition and management of type 2 diabetes. Molecules 2016, 21, 1374, 10.3390/molecules21101374.

[7]   Lattanzio V. Phenolic compounds: Introduction. Ramawat KG, Me´rillon JM, ed, Natural Products. Berlin/Heidelberg,Germany, Springer-Verlag, 2013, 1543–1580. 10.1007/978-3-642-22144-6_57 10.1007/978-3-642-22144-6_57.

[8]   Mann J, Davidson RS, Hobbs JB, Banthorpe DV, Harborne JB. Natural Products: Their chemistry and biological significance. Longman, Harlow, UK, Longman Scientific & Technical, 1994.

[9]   Geissman TA, Crout DHG. Organic chemistry of secondary plant metabolism. Freeman Cooper, SF. USA, Freeman Cooper and Company, USA 1969.

[10]   Torssell KBG. Natural products chemistry: A mechanistic, biosynthetic and ecological approach. Swedish Pharmaceutical Society. Stockholm, Sweden, Swedish Pharmaceutical Press, 1997.

[11]   Brodniewicz T, Grynkiewicz G. Plant phenolics as drug leads-What is missing?. Acta Pol Pharm 2012, 69(6), 1203–1217.

[12]   Halliwell B. Oxidative stress and cancer: Have we moved forward?. Biochem J 2007, 401(1), 1–11.

[13]   Rahman I, Biswass SK, Kirkham PA. Regulation of inflammation and redox signaling by dietary polyphenols. Biochem Pharmacol 2006, 72, 1439–1452.

[14]   Pereira DM, Valentão P, Pereira JA, Andrade PB. Phenolics: From Chemistry to Biology. Molecules 2009, 14(6), 2202–2211.

[15]   Pandey KB, Rizvi SI. Plant polyphenols as dietary antioxidants in human health and disease. Oxid Med Cell Longev 2009, 2(5), 270–278.

[16]   Basli A, Belkacem N, Amrani I. Heath benefits of phenolic compounds against cancers. Soto-Hernandez M, Palma-Tenango M, Garcia-Mateos MR, ed, Phenolic compounds – Biological activity. IntechOpen 2017, 193–210, 10.5772/67232.

[17]   Dai J, Mumper RJ. Plant phenolics: Extraction, analysis and their antioxidant and anticancer properties. Molecules 2010, 15, 7313–7352.

[18]   Lin D, Xiao M, Zhao J, Li Z, Xing B, Li X, Kong M, Li L, Zhang Q, Liu Y, Chen H, Qin W, Wu H, Chen S. An overview of plant phenolic compounds and their importance in human nutrition and management of type 2 diabetes. Molecules 2016, 21, E1374, 10.3390/molecules21101374.

[19] Asgar A. Anti-diabetic potential of phenolic compounds: A review. Int J Food Prop 2013, 16, 91–103.

[20] González-Sarrías A, Núñez-Sánchez MÁ, Tomás-Barberán FA, Espín JC. Neuroprotective Effects of bioavailable polyphenol-derived metabolites against oxidative stress-induced cytotoxicity in human neuroblastoma SH-SY5Y Cells. J Agric Food Chem 2017, 65(4), 752–758.

[21] Basli A, Soulet S, Chaher N, Mérillon JM, Chibane M, Monti JP, Richard T. Wine polyphenols: Potential agents in neuroprotection. Oxid Med Cell Longev 2012, 2012, 805762, 10.1155/2012/805762.

[22] Ozcan T, Akpinar-Bayizit A, Yilmaz-Ersan L, Delikanli B. Phenolics in human health. Int J Chem Eng Appl 2014, 5(5), 393–396.

[23] Dodig S, Čepelak I, Pavić I. Hallmarks of senescence and aging. Biochem Med (Zagreb) 2019, 29(3), 030501, 10.11613/BM.2019.030501.

[24] Lee KE, Bharadwaj S, Yadava U, Kang SG. (2019) Evaluation of caffeine as inhibitor against collagenase, elastase and tyrosinase using *in silico* and *in vitro* approach. J Enzyme Inhib Med Chem 2019, 34(1), 927–936.

[25] Godos J, Sinatra D, Blanco I, Mulè S, La Verde M, Marranzano M. Association between dietary phenolic acids and hypertension in a Mediterranean cohort. Nutrients 2017, 9(10), 1069, 10.3390/nu9101069.

[26] Miranda AM, Steluti J, Fisberg RM, Marchioni DM. Association between polyphenol intake and hypertension in adults and older adults: A population-based study in Brazil. PLoS One 2016, 11(10), e0165791, 10.1371/journal.pone.0165791.

[27] Fehér J, Lengyel G. Silymarin in the prevention and treatment of liver diseases and primary liver cancer. Curr Pharm Biotechnol 2012, 13(1), 210–217.

[28] Nichols JA, Katiyar SK. Skin photoprotection by natural polyphenols: Anti-inflammatory, antioxidant and DNA repair mechanisms. Arch Dermatol Res 2010, 302(2), 71–83.

[29] Minozzo BR, Lemes BM, Justo ADS, Lara JE, Petry VEK, Fernandes D, Belló C, Vellosa JCR, Campagnoli EB, Nunes OC, Kitagawa RR, Avula B, Khan IA, Beltrame FL. Anti-ulcer mechanisms of polyphenol extract of *Euphorbia umbellata* (Pax) Bruyns. J Ethnopharmacol 2016, 191, 29–40.

Kanchana Wijesekera, Aruna S. Dissanayake

4 Terpenes

4.1 Introduction

Terpenes are considered as one of the largest and structurally diverse groups of natural products, found in plants, fungi, and even in humans. They have been used for over two thousand years as fragrances [1]. The principal volatile component of essential oils is often a terpene. Some terpenes and compounds derived from terpenes (known as terpenoids) are used in Western medicine to cure various illnesses. The anticancer drug paclitaxel (**1**) was derived from a diterpene obtained from the Pacific yew tree *Taxus brevifolia* (Figure 4.1).

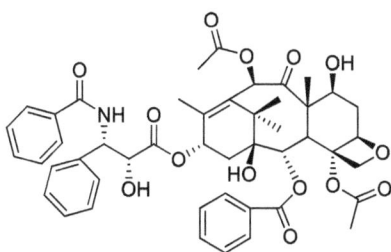

Paclitaxel (**1**)

Figure 4.1: Structure of paclitaxel.

The name "terpene" is derived from the odorous hydrocarbon found in turpentine, the essential component of which is α-pinene, indicating the presence of double bonds with the suffix "ene" [2]. Terpenes are sometimes referred to as isoprenes based on the monomeric building material "isoprene" (repeating C_5 unit), and this is known as the isoprene rule. According to the correct definition, terpenes are pure hydrocarbons of the general formula $(C_5H_8)_n$, and when additional functional groups are present they are known as terpenoids or isoprenoids [3]. Very often the term "terpene" referred not only to hydrocarbons but also to functionally substituted derivatives as well. Most of the terpenoids (isoprenoids) are multicyclic structures with oxygen-containing functional groups. Thousands of terpenoids with a vast diversity of structures are being produced by plants, and they contribute to about 60% of known natural products. In the following structures, the constituent isoprene units are indicated by dotted lines (Figure 4.2).

Kanchana Wijesekera, Faculty of Allied Health Sciences, University of Ruhuna, Galle, Sri Lanka
Aruna S. Dissanayake, Faculty of Allied Health Sciences,University of Ruhuna,Galle, Sri Lanka,
e-mail: kdwijesekera@gmail.com

https://doi.org/10.1515/9783110595949-004

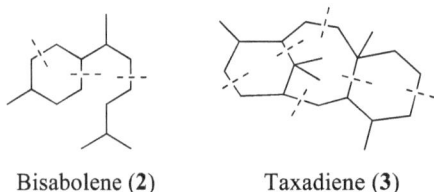

Bisabolene (**2**)   Taxadiene (**3**)

Figure 4.2: Isoprene units of some terpenes.

Terpenes are classified according to the number of the building blocks "isoprene" which is 2-methylbuta-1,3-diene (C_5H_8) units. Usually, these isoprene units are fused together in a head-to-tail fashion and deviation could be seen in some compounds leading to linear or cyclic structures (Figure 4.3). For instance, tail-to-tail linkages occur in higher terpenes such as squalene which is a triterpene ($C_{30}H_{48}$).

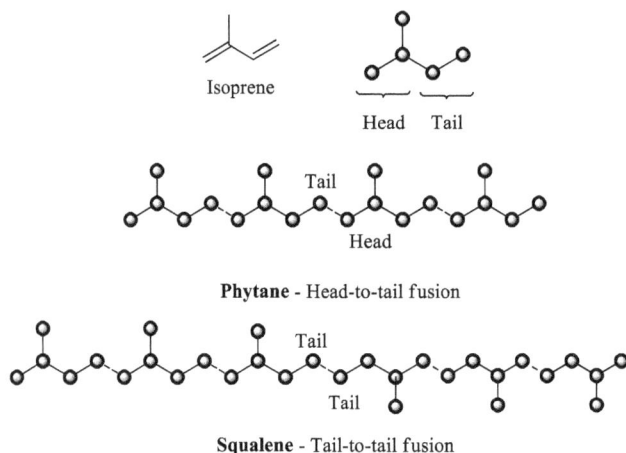

Isoprene

Head   Tail

Tail

Head

Phytane - Head-to-tail fusion

Tail

Tail

Squalene - Tail-to-tail fusion

Figure 4.3: Different fusion patterns of terpenes.

Based on the number of isoprene units present, terpenes are classified as hemi (C_5), mono (C_{10}), sesqui (C_{15}), di (C_{20}), ses (C_{25}), tri (C_{30}), tetra (C_{40}), and poly ($>C_{40}$) terpenes (Table 4.1).

Terpenes are biosynthesized by the reactions between reactive isoprene units (C_5), isopentenyl pyrophosphate (IPP) and dimethylallyl pyrophosphate (DMAPP) (Figure 4.4) which in turn are derived either by mevalonic acid (MVA) or 1-deoxy-D-xylulose-5-phosphate (deoxyxylulose phosphate; DXP) pathways. IPP is enzymatically converted to its double bond isomer DMAPP by IPP isomerase.

Table 4.1: Classification of Terpenes.

No. of isoprene units	General formula	Name	Example
1	C_5H_8	Hemiterpenes	Isovaleric acid
2	$C_{10}H_{16}$	Monoterpenes	Geraniol
3	$C_{15}H_{24}$	Sesquiterpenes	Artemisinin
4	$C_{20}H_{32}$	Diterpenes	Taxol
5	$C_{25}H_{40}$	Sesterpenes	Manoalide
6	$C_{30}H_{48}$	Triterpenes	Squalene
8	$C_{40}H_{64}$	Tetraterpenes	Lycopene
>8	$(C_5H_8)n$	Polyterpenes	Natural rubber

IPP   DMAPP

Figure 4.4: Structures of IPP and DMAPP.

These two compounds are reactive isoprene units. They undergo condensation in a head-to-tail fashion producing geranyl pyrophosphate, a monoterpene containing ten carbon atoms (C_{10}). Hydrolysis of geranyl pyrophosphate gives geraniol, a major constituent of geranium oil and rose oil.

Further condensation of geranyl pyrophosphate with isopentenyl pyrophosphate yields the sesquiterpene (C_{15}) farnesyl pyrophosphate. A repeat of this condensation gives the diterpene (C_{20}). The triterpenes (C_{30}) are not produced by further condensation of diterpenes with IPP, but rather by a reductive coupling (dimerization) of two farnesyl units to give the all-important chemical squalene (Figure 4.5).

The two most important diterpenes are vitamin A (**4**) and abietic acid (**5**). The molecular arrangement of isoprene units in tetraterpenes results in symmetrical compounds with an extended conjugated double-bond system, which acts as light-absorbing chromophore accounts for the yellow, orange, or red color. β-Carotene (**6**) is such a tetraterpene that serves as a precursor of vitamin A (**4**) (Figure 4.6).

The huge diversity of terpenes is due to the fact that they can exhibit different types of isomerism and have a wide range of functional groups (Figure 4.7).

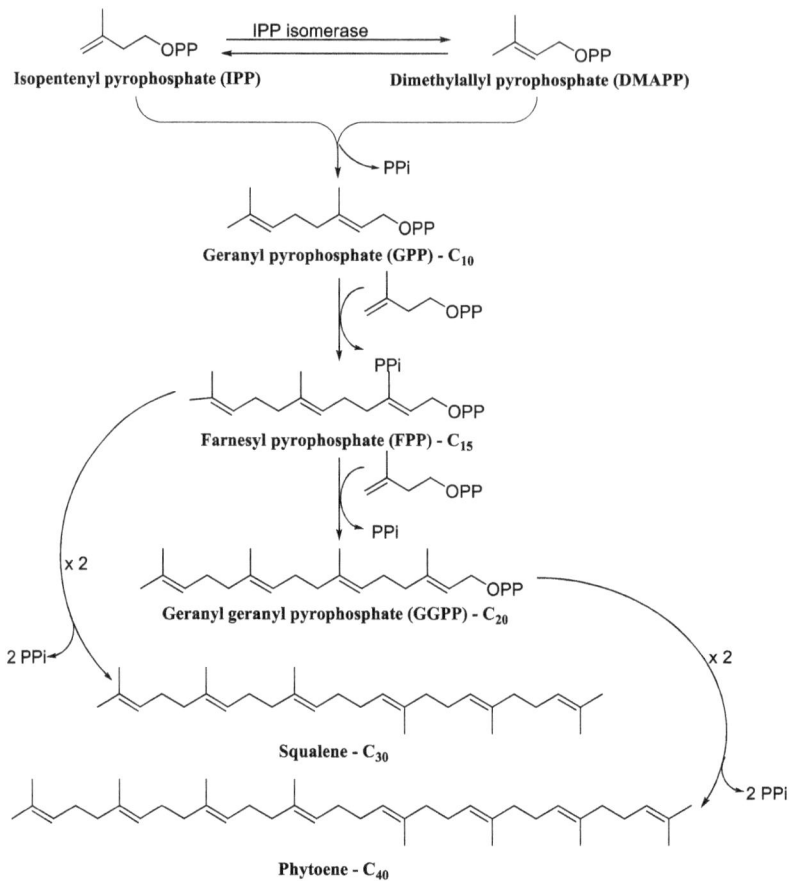

Figure 4.5: Biosynthesis of terpenes.

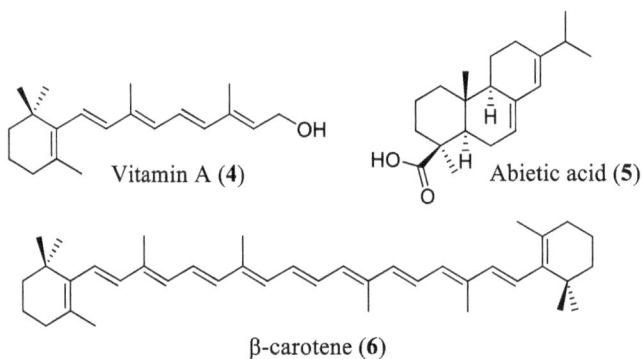

Figure 4.6: The structures of some representative terpenes.

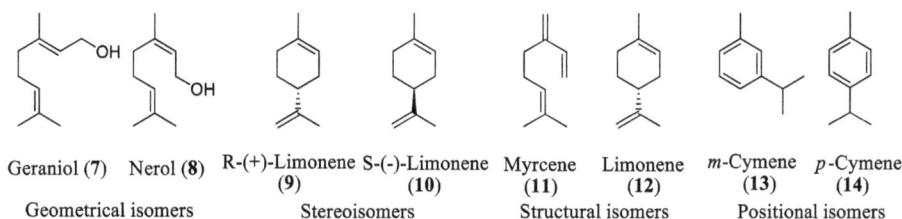

Geraniol (7) Nerol (8) R-(+)-Limonene (9) S-(-)-Limonene (10) Myrcene (11) Limonene (12) *m*-Cymene (13) *p*-Cymene (14)

Geometrical isomers Stereoisomers Structural isomers Positional isomers

Figure 4.7: Examples for isomerism exhibited by terpenes.

4.2 Classification of terpenes

4.2.1 Hemiterpenes (C$_5$)

This is the simplest of all terpenes with a general formula of C$_5$H$_8$. Hemiterpenes are produced via the modifications in dimethylallyl pyrophosphate (DMAPP) and isopentenyl pyrophosphate (IPP). The simplest hemiterpene, isoprene is released by the leaves of many trees and herbs [2]. Tiglic (15) and angelic acids (16) belong to the group of hemiterpenoids and form esters with a number of natural products. In addition, numerous C$_5$ compounds contain the isopentane skeleton, including α-furoic acid (17), β-furoic acid (18), isoamyl alcohol (19), senecioic acid (20), and iso-valeric acid (21) (Figure 4.8). Hemiterpenes are useful in plant defense as they repel herbivores or attract predators and parasites of herbivores [4].

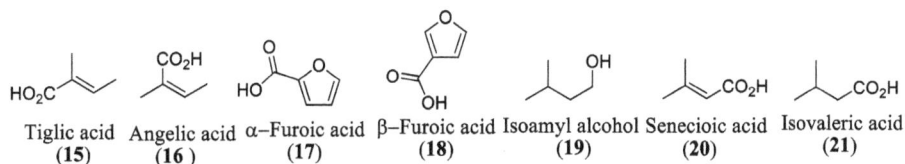

Tiglic acid (15) Angelic acid (16) α-Furoic acid (17) β-Furoic acid (18) Isoamyl alcohol (19) Senecioic acid (20) Isovaleric acid (21)

Figure 4.8: Structures of hemiterpenoids.

4.2.2 Monoterpenes (C$_{10}$)

This is the type of terpene with a general formula of C$_{10}$H$_{16}$ and compounds belonging to this group have characteristic odors and are responsible for the aroma of corresponding plants. Due to this distinguishing odor, most of the members of this family are extensively used in cosmeceutical and food industries and some members are used as pharmaceutical aids in medicine.

Based on the structural arrangement, monoterpenes are classified as acyclic, monocyclic, and bicyclic compounds and further classified according to the ring size. The enormous diversity of monoterpenes is due to the presence of functional groups such as hydroxyl, aldehyde, or ketone in the above basic categories and these are responsible for the distinctive odors and biological activities of the compounds. For example, the monoterpene myrcene, which is used as a flavoring agent, is an olefinic hydrocarbon obtained from hop oil. Geraniol (**7**), citronellal, lavandulol, and linalool are other examples of acyclic monoterpenes. Menthol (**34**) is a well-known example of a monocyclic monoterpene and camphor (**35**) is an example of a bicyclic monoterpene. Some examples of monoterpenes and their uses are given in Table 4.2.

Table 4.2: Monoterpenes: origin and uses.

Type	Acyclic monoterpenes
Myrcene ($C_{10}H_{16}$)	This occurs in bay leaves (*Laurus nobilis*) as well as hops (*Humulus lupulus*) [5]. It is an important intermediate in the manufacture of perfumes.
Ocimene and its isomers ($C_{10}H_{16}$)	This occurs in basil leaves (*Ocimum basilicum*) [5]. It is used in the manufacture of perfumes.
Geraniol ($C_{10}H_{18}O$)	This is obtained from rose oil [6]. It is used in the manufacture of perfumes. It is an important intermediate in the manufacture of geranyl esters such as citronellol and citral.
Citronellal ($C_{10}H_{18}O$)	This is found in the leaves and stems of different species of *Cymbopogon* (lemongrass) [7]. It is used as an insect repellant as well as for the manufacture of incense, perfumes, cosmetics, etc. It is also used as a flavoring agent.

Table 4.2 (continued)

Type	Acyclic monoterpenes
Citral (C$_{10}$H$_{16}$O) Geranial Neral Citral a Citral b	This is isolated from the leaves and stems of different species of *Cymbopogon* (lemongrass) [8]. It is used in the production of perfumes and as a starting material for the synthesis of vitamin A.
Linalool (C$_{10}$H$_{18}$O) 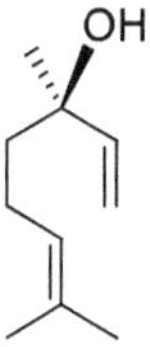	Of the two optical isomers; R (-) form (shown here) occurs in coriander and S (+) form in orange oil [9]. It is used as a fragrance component in perfumes, cosmetics, soaps, and detergents. This is an important synthetic intermediate.
Lavandulol (C$_{10}$H$_{18}$O) 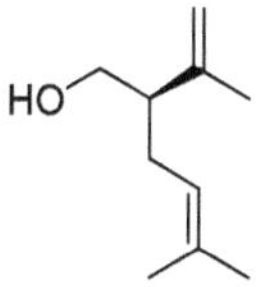	This is obtained from the oil of lavender (*Lavandula augustifolia*) and is commonly used in men's perfumes [10, 11]

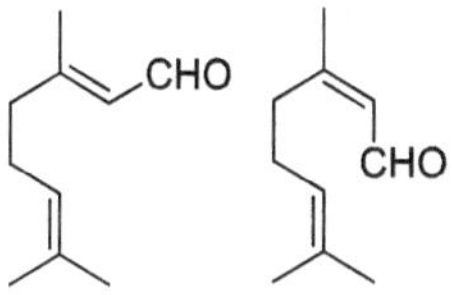

Type	Monocyclic monoterpenes
α-Terpineol (C$_{10}$H$_{18}$O) 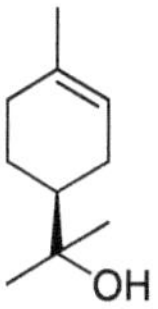	This is found as the main ingredient in *Melaleuca alternifolia* (tea tree) oil. Due to the pleasant odor, it is used widely in the manufacture of perfumes, cosmetics, and aromatic scents [12].
Carvone (C$_{10}$H$_{14}$O) 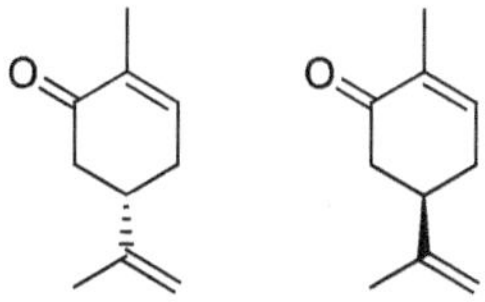 S-(+)-Carvone R-(-)-Carvone	This is obtained from the oil of *Carum carvi* (Caraway) and is an important ingredient in the manufacture of gripe water [13].

Table 4.2 (continued)

Type	Monocyclic monoterpenes
Menthol ($C_{10}H_{20}O$)	This is obtained from the fresh flowering tops of the plants commonly known as *Mentha piperita* or other species of *Mentha* [14]. It is used as an aromatic agent for various types of mouthwashes, toothpaste and similar oral formulations, nasal sprays, and inhalants, as a flavoring agent for chewing gums, candies, throat lozenges, and also certain mentholated cigarettes.
Limonene ($C_{10}H_{16}$) (*R*)-limonene	This is the dominant component of mandarin peel oil from *Citrus reticulata* [15].
Phellandrenes ($C_{10}H_{16}$) α β	Phellandrenes are a pair of organic compounds isolated from *Eucalyptus* spp. The compounds share an analogous molecular structure and chemical properties [16].
Thymol ($C_{10}H_{14}O$)	This is obtained from the essential oil of *Thymus vulgaris* (Thyme oil), *Monarda punctata* (Horsemint oil), and *Monarda didyma* (Oswego tea oil) [5, 17]. It is employed as an antifungal and antibacterial agent and used as a component in several analgesic and topical antiseptic formulations.
Cymenes ($C_{10}H_{14}O$) *m*-cymene *p*-cymene	*m*-Cymene is a constituent of the ethereal oil of blackcurrant. *p*-Cymene occurs in the ethereal oil of cinnamon (*Cinnamomum zeylanicum*), eucalyptus (*Eucalyptus globulus*) and nutmeg (*Myristica fragrans*). Cymenes are used as an antimicrobial agent in topical application for symptomatic treatment of common skin disorders, treatment of wounds, and vaginitis [18].

Table 4.2 (continued)

Type	Bicyclic monoterpenes (6 + 3 membered ring)
(+)-3-Carene (C$_{10}$H$_{16}$)	It is obtained from the oil of turpentine (*Pinus longifolia*) and some species of fir (*Abies*), juniper (*Juniperus*), and *Citrus* [5].
(+)-4(10)-Thujene (C$_{10}$H$_{16}$)	(+)-4(10)-Thujene, better known as [(+)]-sabinene is obtained from the fresh tops of *Juniperus sabina* [5].

Type	Bicyclic monoterpenes (6 + 4 membered ring)
α- and β-Pinene (C$_{10}$H$_{16}$) α-pinene β-pinene	This is chiefly obtained from the oil of turpentine from the woods of *Pinus caribaea*, *P. palustris* and *P. pinaster*, etc. [5]. α-pinene is also obtained from juniper oil (*Juniperus communis*) and used as an antiseptic as well as in aromatherapy [19].
(+)-Verbenone (C$_{10}$H$_{14}$O)	This occurs as pheromones of bark beetles *Ips typographus* as well as a constituent of Spanish verbena oil obtained from *Verbena triphylla* [5].

Type	Bicyclic monoterpenes (6 + 5 membered ring)
Camphor (C$_{10}$H$_{16}$O)	This is obtained from the oil from camphor tree (*Cinnamomum camphora*) and used as an antiseptic [19].
Borneol (C$_{10}$H$_{18}$O)	This is found in *Thymus vulgaris*, *T. serpyllum*, and sage (*Salvia officinalis*) [19].

Even though the majority of cyclic monoterpenes are based on either cyclopentane or cyclohexane rings, some extraordinary monoterpenes comprise cyclopropane and cyclobutane ring systems. The esters of chrysanthemic (**22**) and pyrethric acids (**23**); i.e., cinerins (**25**), jasmolins (**26**) and pyrethrins (**27**) and other related compounds with insect repellant activity obtained from the dried flowers of *Chrysanthemum cinerariifolium* and (+)-chrysanthemol (**24**) obtained from the leaves of *Artemisia ludoviciana* are examples for cyclopropane monoterpenes (Figure 4.9) [5].

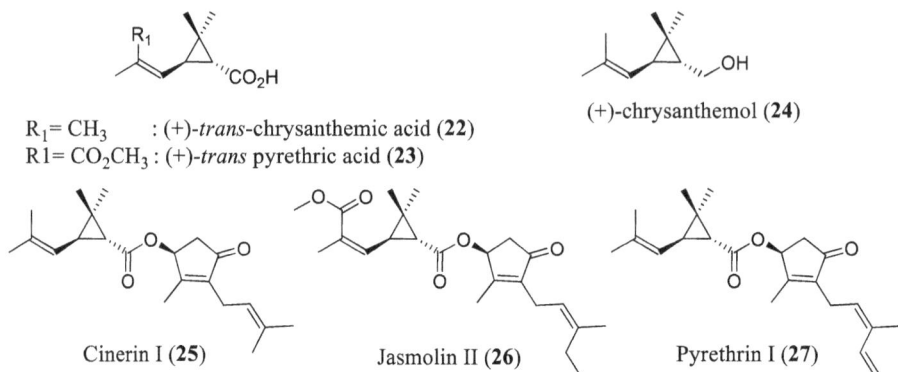

R₁= CH₃ : (+)-*trans*-chrysanthemic acid (**22**)
R1= CO₂CH₃ : (+)-*trans* pyrethric acid (**23**)

(+)-chrysanthemol (**24**)

Cinerin I (**25**)   Jasmolin II (**26**)   Pyrethrin I (**27**)

Figure 4.9: Examples for cyclopropane monoterpenoids.

Grandisol (**28**), a cyclobutane monoterpene, is the major sexual pheromone of the agricultural pest "cotton boll weevil" (*Anthonomus grandis*). Other cyclobutane monoterpenes include junionone (**29**) from the juniper tree *Juniperus communis* and fragranol (**30**) from *Artemisia fragrans* (Figure 4.10) [5].

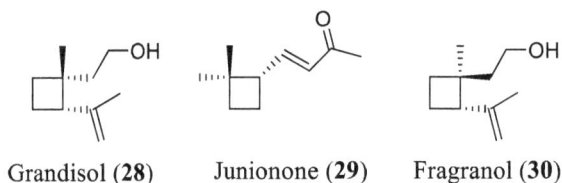

Grandisol (**28**)   Junionone (**29**)   Fragranol (**30**)

Figure 4.10: Examples for cyclobutane monoterpenoids.

Iridoids represent a group of cyclopentane-pyran monoterpenes and are found as natural constituents in a large number of plant families. Compounds belonging to this group are highly oxygenated. The name *iridoid* is a generic name derived from several compounds isolated from a genus of ants *Iridomyrmex* [20]. For example, iridomyrmecin (**31**) is an insecticidal pheromone of the ant *Iridomyrmex humilis* and nepetalactones (**32**) from the volatile oil of catnip (*Nepeta cataria*) (Figure 4.11), which strongly attracts cats, belong to iridoids [5, 21].

Monoterpenes are biosynthesized by the reaction between DMAPP and IPP which ultimately generates a C_{10} unit, geranyl pyrophosphate. Different types of monoterpenes including simple acyclic monoterpenes such as myrcene, linalool, and, cyclic monoterpenes including menthol, camphor, thymol, etc. are generated

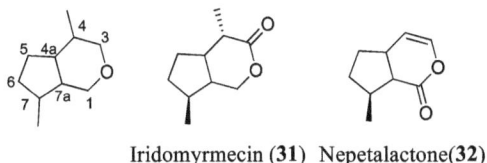

Iridomyrmecin (**31**) Nepetalactone(**32**)

Figure 4.11: Terpenes with of iridoid skeleton.

from geranyl pyrophosphate [22, 23] (Figure 4.12). Even though the monoterpenes contain only two isoprene units, different functional groups and the presence of chiral centers had created an amazing diversity in this group.

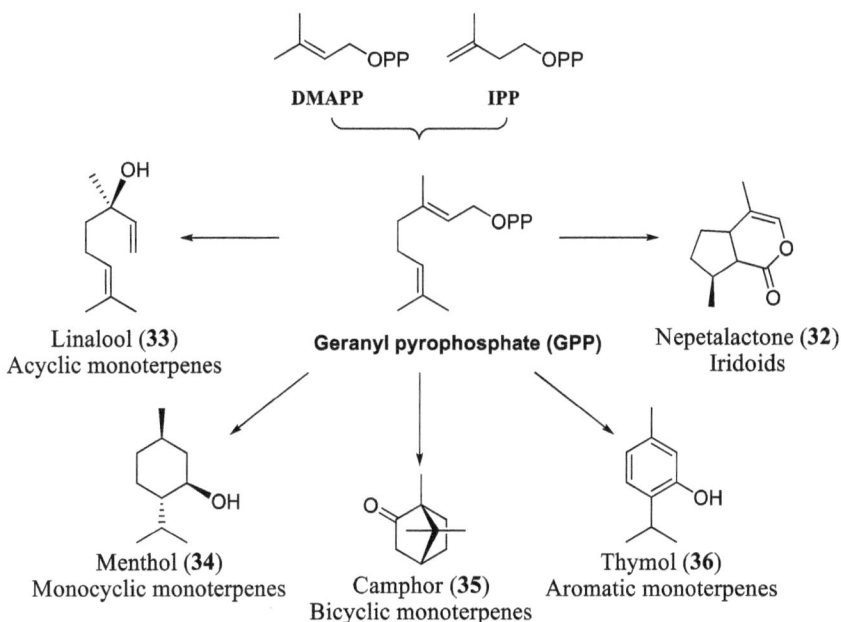

Figure 4.12: Formation of monoterpenes.

4.2.3 Sesquiterpenes (C$_{15}$)

Compounds belonging to this group have a general formula of $C_{15}H_{24}$ and show properties that are similar to monoterpenes. Sesquiterpenes show greater potential for stereochemical diversity and possess stronger odors. It is reported that the compounds belonging to this group show antimicrobial, anti-insecticidal, and anti-inflammatory properties (Table 4.3). Hence, sesquiterpenes are important as chemical protectors for producing organisms [19, 24]. Structurally, sesquiterpenes

are derived from farnesyl pyrophosphate (FPP), which is composed of three isoprene units and shows acyclic, monocyclic, bicyclic, and tricyclic arrangements.
A few typical examples are shown in Figure 4.13.

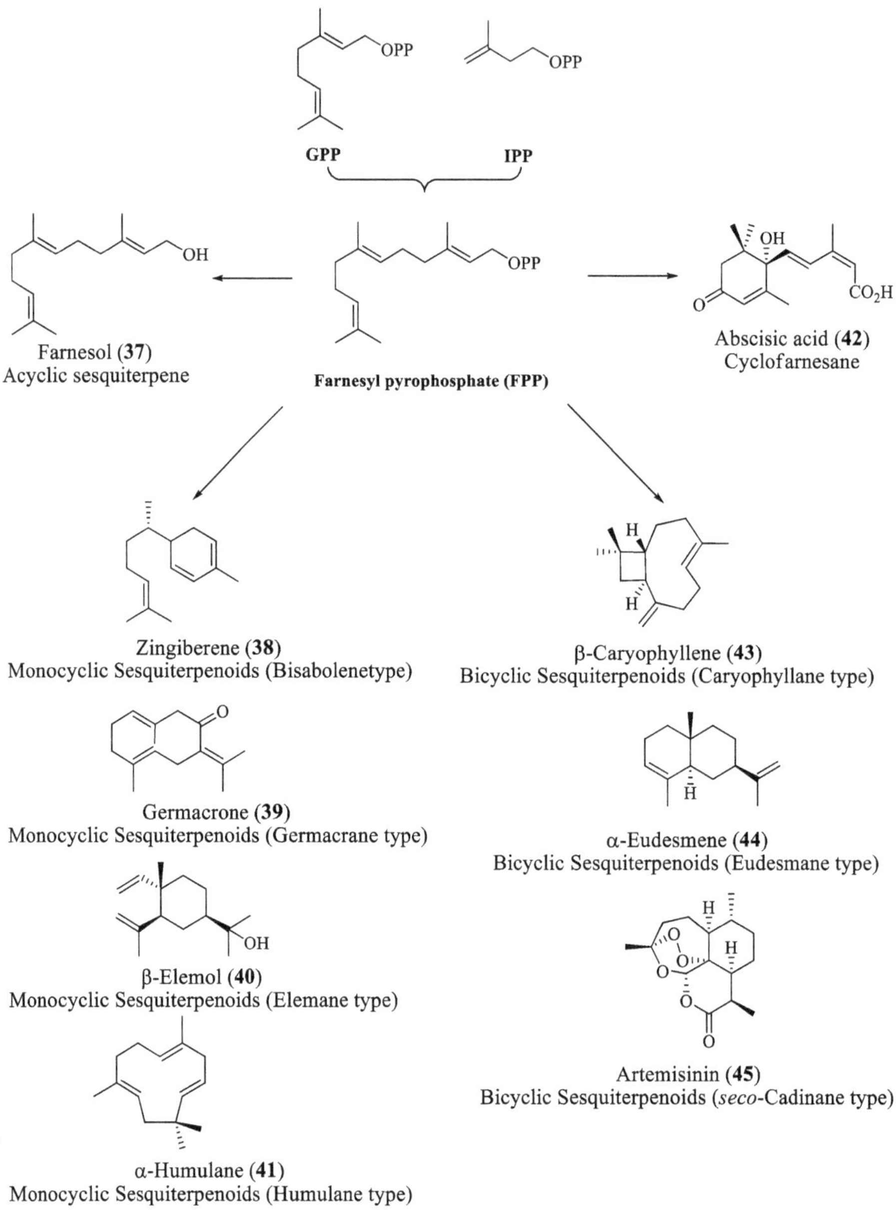

Figure 4.13: Formation of sesquiterpenes.

Some members of the sesquiterpenes are pharmaceutically important. Artemisinin from *Artemisia annua* is a potent anti-malarial drug. Avarone is a sesquiterpene hydroquinone isolated from marine sponge *Dysidea avara* and exhibited potential antiviral activity against HIV.

Sesquiterpene lactones are considered as a separate group due to unique chemical and biological activities. The noticeable feature of all the compounds belonging to this group is the presence of γ-lactone moiety. Over 6,000 compounds of this group are known and divided into three major classes as guaianolides, germacranolides, and eudesmanolides (Figure 4.14) [25–27].

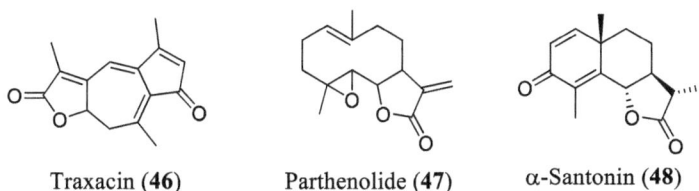

Traxacin (**46**)   Parthenolide (**47**)   α-Santonin (**48**)

Figure 4.14: Sesquiterpene lactones Taraxacin – a guaianolide, Parthenolide – a germacranolide, α-Santonin – an eudesmanolide.

Table 4.3: Sesquiterpenes: origin and uses.

Type	Acyclic sesquiterpenes
Farnesol ($C_{15}H_{26}O$) 	This is an important intermediate in terpene biosynthesis and is found in many essential oils. It is reported with cancer chemopreventive activity [28].
Nerolidol ($C_{15}H_{26}O$) 	Mainly obtained from oil of neroli and flower oils of jasmine, rose, *Hibiscus*, etc. and is extensively used as an ingredient in perfumery [5, 29].
Dendrolasin ($C_{15}H_{22}O$) 	This is a furan containing sesquiterpene isolated from marine snails and sweet potatoes. It acts as an alarm and defense pheromone of the ant *Dendrolasius fuliginosus* [5].

Table 4.3 (continued)

Type	Monocyclic sesquiterpenes
Zingiberene (C$_{15}$H$_{24}$)	This is a bisobolene-type sesquiterpene obtained from the oil of ginger (*Zingiber officinale*). Zingiberene is responsible for the characteristic flavor of ginger [30].
Abscisic acid (C$_{15}$H$_{20}$O$_4$)	This is a cyclofarnesane found in many plants. Abscisic acid acts as a plant growth regulator that controls the flowering, shedding of leaves, etc., of plants [31].
Germncrene C (C$_{15}$H$_{24}$)	Germacrene C is one of the germacrene-type sesquiterpenes isolated from the citrus peel oils [32].
β-Elemenone (C$_{15}$H$_{22}$O)	This is a sesquiterpenoid with a structure based on the elemane skeleton. Elemane is a monocyclic compound. This is an elemane-type sesquiterpene isolated from the oil of myrrh [33].
α- Humulene (C$_{15}$H$_{24}$)	This is a humulane-type sesquiterpene isolated from the leaves of *Lindera strychnifolia* [34].

Type	Bicyclic sesquiterpenes
β-Caryophyllene (C$_{15}$H$_{24}$)	This is found in the oil obtained from clove (*Syzygium aromaticum*) [35].
β-Eudesmol (C$_{15}$H$_{26}$O) β-eudesmol	β-eudesmol has been found in *Dioscorea japonica* and exhibits antimutagenic activity [36].

Table 4.3 (continued)

Type	Bicyclic sesquiterpenes
Artemisinin ($C_{15}H_{22}O_5$) 	This is an unusual sesquiterpene with an endoperoxide moiety and is isolated from the Chinese ornamental plant sweet wormwood (*Artemisia annua*). Artemisinin is successful in treating complicated species of *Plasmodium falciparum* [37].

Type	Tricyclic sesquiterpenes
Paesslerins A ($C_{17}H_{26}O_2$) and B ($C_{19}H_{28}O_4$) Paesslerins A Paesslerins B	These two cytotoxic sesquiterpenoids are isolated from the marine sponge *Alcyonium paessleri* [38].

4.2.4 Diterpenes (C_{20})

Compounds belonging to this group have a general formula of $C_{20}H_{32}$ with four isoprene units. Due to higher boiling points, compounds in this group are not considered as essential oils. The best-selling anticancer drug "paclitaxel (**1**)" from *Taxus brevifolia*, the inimitable chemical compounds "ginkgolides" from *Ginkgo biloba* belong to the group of diterpenes. Diterpenes are derived by geranylgeranyl pyrophosphate (GGPP) which forms by the reaction between farnesyl pyrophosphate (FPP, C_{15}) with isopentenyl pyrophosphate (IPP, C_5) (Figure 4.15). Based on the cyclization and spatial arrangements, thousands of diterpenes are being formed in plants, fungi, and marine organisms, etc. Diterpenes exhibit acyclic, monocyclic, bicyclic, tricyclic, and tetracyclic structures and possess various types of biological activities such as anticancer, antimycobacterial, antimalarial, and antifungal (Table 4.4) [39–41].

4.2.5 Sesterpenes (C_{25})

Compounds belonging to this group have a general formula of $C_{25}H_{40}$ with five isoprene units. Several sesterpenoids are known, and they are predominantly found in fungi and marine organisms. Sesterpenes are derived by geranylfarnesyl pyrophosphate (GFPP) which is formed by the reaction between geranylgeranyl pyrophosphate

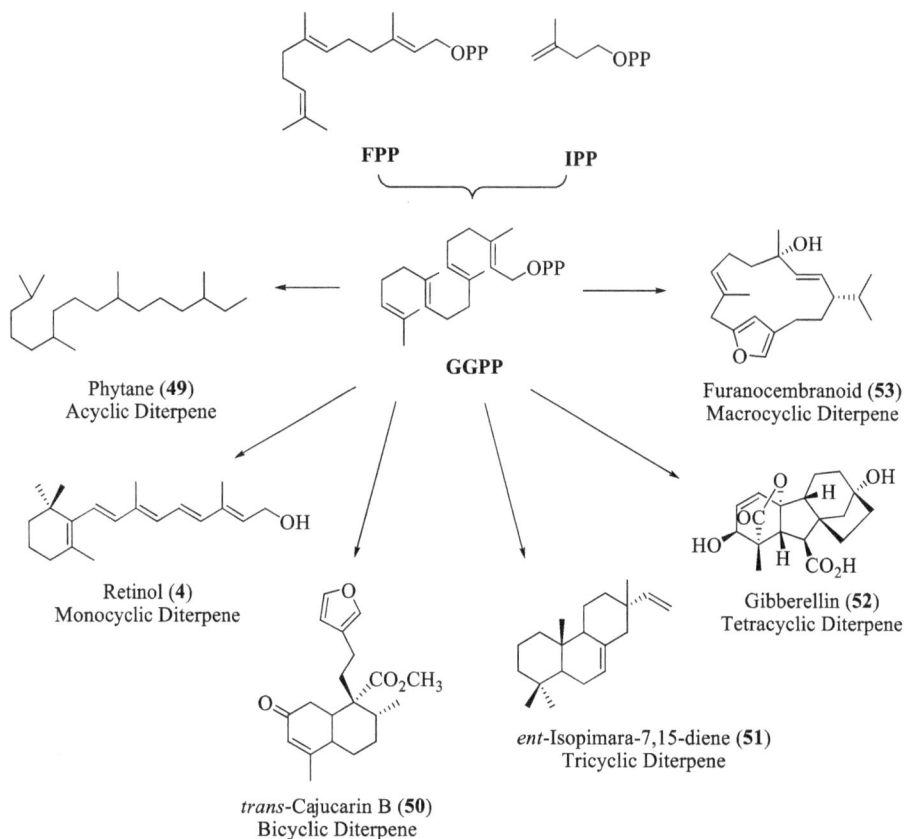

Figure 4.15: Formation of diterpenes.

Table 4.4: Different types of diterpenes and their origins.

Type	Acyclic diterpenes
Phytol ($C_{20}H_{40}O$)	Phytol, the prenyl side chain of chlorophyll, is an acyclic diterpene and possesses various biological activities [42, 43].
Type	**Monocyclic diterpenes**
Retinol ($C_{20}H_{30}O$)	This is also known as vitamin A_1 is a key substance in the visual process and is important for human development, etc. It is found in many foods including carrot, egg, cod liver oil, etc.

Table 4.4 (continued)

Type	Monocyclic diterpenes
Retinal ($C_{20}H_{28}O$)	This is the vitamin A aldehyde that binds to the apoprotein opsin.

Type	Bicyclic diterpenes
Nasimalun A ($C_{21}H_{24}O$)	This is a clerodane-type diterpene isolated from the roots of *Barringtonia racemosa* [44, 45].
Crotohalimaneic acid ($C_{20}H_{28}O_3$)	This is a halimane-type diterpene isolated from the stem barks of *Croton oblongifolius* and exhibits strong cytotoxicity against a panel of human tumor cell lines [46].
Labda-7,12(E),14-triene ($C_{20}H_{32}$)	This is a labdane-type diterpene isolated from the stem barks of *Croton oblongifolius* [47].

Type	Tricyclic diterpenes
Diaporthein A ($C_{20}H_{30}O_6$)	This is a pimarane-type diterpene from the culture broth of the fungus *Diaporthe* spp. [48].

Table 4.4 (continued)

Type	Tricyclic diterpenes
Podocarpinol ($C_{17}H_{24}O_2$)	Podocarpinol is a tricyclic diterpene obtained from the heartwood of *Podocarpus totara* [49].

Type	Tetracyclic diterpenes
Gibberellic acid ($C_{19}H_{22}O_6$)	This is a tetracyclic diterpene isolated from the fungus *Gibberella fujikuroi* and it serves as a plant growth regulator [50].
Scopadulin ($C_{27}H_{38}O_5$)	This is an unusual tetracyclic diterpene obtained from the flowering plant *Scoparia dulcis* [51].

Type	Macrocyclic diterpenes
Crotocembraneic acid ($C_{20}H_{30}O_2$)	Crotocembraneic acid and other related cembranoids were isolated from the stem barks of the Thai medicinal plant *Croton oblongifolius* [52, 53].
10-Oxodepressin ($C_{20}H_{28}O_2$)	This is a casbane-type diterpene obtained from the leaves of *Oryza sativa* [54].

Table 4.4 (continued)

Type	Macrocyclic diterpenes
Taxol ($C_{47}H_{51}NO_{14}$)	This is an unusual diterpene isolated from the bark of the Pacific yew, *Taxus brevifolia* with a potent antileukemic and tumor inhibitory properties [55].

Type	Other important diterpenes
Ginkgolides A ($C_{20}H_{24}O_9$)	Ginkgolides A and other related diterpenic lactones were isolated from the leaves of *Ginkgo biloba* [56].
Solenolide A ($C_{28}H_{41}ClO_9$)	This is an antiviral diterpene isolated from the marine octocoral of the genus *Solenopodium* [57].

(GGPP, C20) with isopentenyl pyrophosphate (IPP, C5) and exist as acyclic, monocyclic, bicyclic, tricyclic, tetracyclic, and pentacyclic structures (Table 4.5).

Table 4.5: Sesterpenes and their origin.

Type	Acyclic sesterpenes
Ircinin I ($C_{25}H_{30}O_5$)	Acyclic sesterterpenes bridged by furan rings have been found in marine organisms. Ircinin I was isolated from various marine sponges and it exhibited antibacterial activity [58].

Table 4.5 (continued)

Type	Monocyclic sesterpenes
Manoalide ($C_{25}H_{36}O_5$)	Manoalide is a sesterpene lactone isolated from the marine sponge *Luffariella variabilis* [59].

Type	Bicyclic sesterpenes
Bilosespen A ($C_{25}H_{40}O_2$)	This is a bicyclic cytotoxic sesterpene isolated from the marine Sponge *Dysidea cinerea* [60].

Type	Tricyclic sesterpenes
Cheilanthane I ($C_{25}H_{36}O_3$)	This is a tricyclic sesterpenoid isolated from a marine sponge of the genus *Ircinia* and exhibited protein kinase inhibitory activity [61].

Type	Tetracyclic sesterpenes
Desoxyscalarin ($C_{25}H_{40}O_3$)	This is a tetracyclic sesterpene isolated from *Spongia officinalis* [5].

Type	Miscellaneous sesterpenes
Bolivianine ($C_{25}H_{30}O_3$)	This is a sesterpene with an unusual skeleton isolated from the plant *Hedyosmum angustifolium* [62].

4.2.6 Triterpenes (C$_{30}$)

Compounds belonging to this group have a general formula of C$_{30}$H$_{48}$ with six isoprene units. Not like the terpenes discussed so far, triterpenes are biosynthesized by the reductive coupling (dimerization) of two farnesyl pyrophosphate (FPP) units in tail-to-tail manner (Figure 4.16). The resulting product is squalene which was originally isolated from the liver oil of shark (*Squalus* sp.) and exists as acyclic, tetracyclic, and pentacyclic structures with oxygenated functional groups such as alcohols and carboxylic acids. The squalene structure is usually drawn in a folded form (Table 4.6) to indicate the ease with which it can undergo cyclization to the steroid nucleus. Cholesterol is one of the simplest, but an important, members of tetracyclic triterpenes. In addition, the aglycone portion of some saponins are also made up of pentacyclic triterpenes.

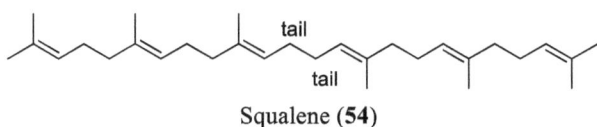

Squalene (**54**)

Figure 4.16: Tail-to-tail assembly of squalene.

Table 4.6: Different types of triterpenes and their origins.

Type	Acyclic triterpenes
Squalene (C$_{30}$H$_{50}$)	This is a highly unsaturated triterpene isolated from the shark liver oil [63].

Type	Tetracyclic triterpenes
Fusidic acid (C$_{31}$H$_{48}$O$_6$)	This is a triterpene isolated from the fermentation broth of *Fusidium coccineum* [64]. It possesses antibacterial activity and is used to treat bacterial infections and available as creams, ointments, etc.

Table 4.6 (continued)

Type	Tetracyclic triterpenes
Lanosterol ($C_{30}H_{50}O$)	Lanosterol is a typical animal triterpenoid which is the precursor of cholesterol and another sterol biosynthesis in animals [65, 66].
Cucurbitacin E ($C_{32}H_{44}O_8$)	Cucurbitacin E and other compounds belonging to this group are highly oxygenated tetracyclic triterpenes. These are found in cucumber, melon (Family Cucurbitaceae), etc. and are bitter tasting [65, 67]. This compound exhibited anti-inflammatory activity by selectively inhibiting COX-2 enzyme [68].
Quassin ($C_{22}H_{28}O_6$)	Quassin is one of the quassinoids isolated from *Quassia amara* and considered as one of the most bitter substances found in nature [37]. This compound is considered as a triterpene lactone.

Type	Pentacyclic triterpenes
β-Boswellic acid ($C_{30}H_{48}O_3$)	This is one of the pentacyclic boswellic acids isolated from the resin of *Boswellia serrata*. Boswellic acids were found to exhibit anti-inflammatory activity [69, 70].

Table 4.6 (continued)

Type	Pentacyclic triterpenes
β-Amyrin (C$_{30}$H$_{50}$O)	β-Amyrin is one of the pentacyclic triterpene isolated from the stem bark resin of *Protium heptaphyllum* [71].
Glycyrrhetinic acid (C$_{30}$H$_{46}$O$_4$)	Glycyrrhetinic acid is the aglycone part of glycyrrhizin which is considered as a saponin glycoside isolated from the roots of *Glycyrrhiza glabra* [65].
Quillaic acid (C$_{30}$H$_{46}$O$_5$)	Quillaic acid is the principal aglycone found in the saponin mixture of Quillaia (*Quillaja saponaria*) [72].
Betulinic acid (C$_{30}$H$_{48}$O$_3$)	This pentacyclic triterpenic acid is isolated from the bark of white birch (*Betula pubescens*) [73]. Betulinic acid exhibited various biological activities, including antiviral, anticancer, etc. [74, 75].

4.2.7 Tetraterpenes (C_{40})

Compounds belonging to this group have a general formula of $C_{40}H_{64}$ with eight isoprene units. Similar to triterpenes, they are formed by the combination of two molecules of geranylgeranyl pyrophosphate (C_{20}) in a tail-to-tail manner. Teteraterpenes are sometimes referred to as carotenoids, where they occur in nature as orange or yellow-colored pigments (Table 4.7). The orange color of carrots and red color of tomatoes are due to β-carotene (**6**) and lycopene, respectively. Together with chlorophyll,

Table 4.7: Different types of tetraterpenes and their origins.

Type	Acyclic triterpenes
Lycopene ($C_{40}H_{56}$)	Lycopene is the bright red colored pigment found in tomatoes (*Solanum lycopersicum*). This is an open-chain hydrocarbon containing 11 conjugated and 2 non-conjugated double bonds arranged in a linear array.
β-carotene ($C_{40}H_{56}$)	This is the colored pigment found in carrots (*Daucus carota*). Enzymes in the body cleave it to a diterpene, vitamin A.
Capsanthin ($C_{40}H_{56}O_3$)	The bright red color of ripen pepper (*Capsicum annuum*) is due to capsanthin.
Zeaxanthin ($C_{40}H_{56}O_2$)	This is one of the macula pigments which are important in vision [77]. It acts as "internal sunglasses"
Capsorubin ($C_{40}H_{56}O_4$)	The ripe fruit of pepper (*Capsicum annuum*) is a rich source of capsorubin [78, 79].

carotenes play a major part in the photosynthesis process of plants as carotenes are highly conjugated natural products that can strongly absorb visible light. The simplest member of this group is lycopene, and different types of compounds are formed when one or both ends of this molecule get cyclized [76].

4.2.8 Polyterpenes ($C_{40>}$)

Compounds belonging to this group contain more than eight isoprene units and have higher molecular weights. One of the well-known examples is natural rubber which contains isoprene units in *cis*-configuration. Gutta-percha is an example of *trans* polyisoprene (Table 4.8).

Table 4.8: Different types of polyterpenes and their origins.

Type	Acyclic triterpenes
Natural rubber (*cis*-1,4-polyisoprene) 	This is an elastic substance obtained from the latex of rubber tree (*Hevea brasiliensis*).
Gutta-percha (*trans*-1,4-polyisoprene) 	This is the major constituent of the milky sap of gutta-percha trees (*Palaquium gutta*)

4.3 Conclusion

Terpenes, terpenoids, or isoprenoids are a major and important class of natural products, and those isolated and characterized so far have numerous valuable pharmaceutical applications. Therefore, it is not surprising that they are processed on a large scale by the chemical and pharmaceutical industries.

References

[1]   Brielmann HL, Setzer WN, Kaufman PB, Kirakosyan A, Cseke LJ. Phytochemicals: The chemical components of plants. Kaufmann PB, Cseke LJ, Warber S, Duke JA, Brielmann HL, eds, Natural products from plants, 2nd Boca Raton. USA, CRC Press Inc, 2006, 1–49.
[2]   McCreath SB, Delgoda R. Pharmacognosy: Fundamentals, applications and strategies. 1st. San Diego, CA, USA, Academic Press, 2017.

[3]   McNaught AD, Wilkinson A, Jenkins A. IUPAC compendium of chemical terminology-the gold book, International Union of Pure and Applied Chemistry, 2006.

[4]   Holopainen JK. Multiple functions of inducible plant volatiles. Trends Plant Sci 2004, 9(11), 529–533.

[5]   Breitmaier E. Terpenes: Flavors, fragrances, pharmaca, pheromones. Wiley-VCH, Weinheim, Germany. Wiley-VCH Verlag GmbH & Co. 2006.

[6]   Chen W, Viljoen AM. Geraniol – a review of a commercially important fragrance material. South African J Bot 2010, 76(4), 643–651.

[7]   Akhila A. Biosynthesis of monoterpenes in *Cymbopogon winterianus*. Phytochemistry 1986, 25(2), 421–424.

[8]   Akhila A. Biosynthetic relationship of citral-trans and citral-cis in *Cymbopogon flexuosus* (lemongrass). Phytochemistry 1985, 24(11), 2585–2587.

[9]   Oliver JE. (S)(+)-Linalool from oil of coriander. J Essent Oil Res 2003, 15(1), 31–33.

[10] Hajhashemi V, Ghannadi A, Sharif B. Anti-inflammatory and analgesic properties of the leaf extracts and essential oil of *Lavandula angustifolia* Mill. J Ethnopharmacol 2003, 89(1), 67–71.

[11] Shellie R, Mondello L, Marriott P, Dugo G. Characterisation of lavender essential oils by using gas chromatography–mass spectrometry with correlation of linear retention indices and comparison with comprehensive two-dimensional gas chromatography. J Chromatogr A 2002, 970(1–2), 225–234.

[12] Khaleel C, Tabanca N, Buchbauer G. α-Terpineol, a natural monoterpene: A review of its biological properties. Open Chem 2018, 16(1), 349–361.

[13] Bouwmeester HJ, Gershenzon J, Konings MC, Croteau R. Biosynthesis of the monoterpenes limonene and carvone in the fruit of caraway. I. Demonstration of enzyme activities and their changes with development. Plant Physiol 1998, 117(3), 901–912.

[14] Moghtader M. *In vitro* antifungal effects of the essential oil of *Mentha piperita* L. and its comparison with synthetic menthol on *Aspergillus niger*. African J Plant Sci 2013, 7(11), 521–527.

[15] Pérez LM, Pauly G, Carde JP, Belingheri L, Gleizes M. Biosynthesis of limonene by isolated chromoplasts from *Citrus sinensis* fruits. Plant Physiol Biochem 1990, 28(2), 221–229.

[16] Lassak EV, Southwell IA. The steam volatile leaf oils of some species of *Eucalyptus* subseries *Strictinae*. Phytochemistry 1982, 21(9), 2257–2261.

[17] Basch E, Ulbricht C, Hammerness P, Bevins A, Sollars D. Thyme (*Thymus vulgaris* L.), thymol. J Herb Pharmacother 2004, 4(1), 49–67.

[18] Marchese A, Arciola CR, Barbieri R, Silva AS, Nabavi SF, Tsetegho Sokeng AJ, Izadi M, Jafari NJ, Suntar I, Daglia M, Nabavi SM. Update on monoterpenes as antimicrobial agents: A particular focus on p-cymene. Materials (Basel) 2017, 10(8), 947. 10.3390/ma10080947.

[19] Heinrich M, Barnes J, Prieto-Garcia J, Gibbons S, Williamson E. Fundamentals of pharmacognosy and phytotherapy 3rd ed. Amsterdam, Netherlands. Elsevier. 2012.

[20] El-Naggar LJ, Beal JL. Iridoids. A Review J Nat Prod 1980, 43(6), 649–707.

[21] Regnier FE, Eeisenbraun EJ, Waller GR. Nepetalactone and epinepetalactone from *Nepeta cataria* L. Phytochemistry 1967, 6(9), 1271–1280.

[22] Turner G, Gershenzon J, Nielson EE, Froehlich JE, Croteau R. Limonene synthase, the enzyme responsible for monoterpene biosynthesis in peppermint, is localized to leucoplasts of oil gland secretory cells. Plant Physiol 1999, 120(3), 879–886.

[23] Chappell J. Biochemistry: Another aspect of nature's ingenuity. Nature 2012, 492(7427), 50–51.

[24] Buckle J. Basic plant taxonomy, basic essential oil chemistry, extraction, biosynthesis, and analysis. Clin Aromather 2015, 37–72.

[25] Evans WC. Treas and evans pharmacognosy. 14th. London, UK, Saunders Company Ltd, 1996.

[26] Ahmad VU, Yasmeen S, Ali Z, Khan MA, Choudhary MI, Akhtar F, Miana GA, Zahid M. Taraxacin, a new guaianolide from *Taraxacum wallichii*. J Nat Prod 2000, 63(7), 1010–1011.

[27] Yang L, Dai J, Sakai J, Ando M. Biotransformation of alpha-santonin by cell suspension cultures of five plants. Biotechnol Lett 2005, 27(11), 793–797.

[28] Crowell P, Gould M. Cancer chemopreventive activity of monoterpenes and other isoprenoids. Kelloff GJ, Hawk ET, Sigman CC, eds, Cancer chemoprevention. Totowa, NJ, USA, Humana Press, 2004, 371–378.

[29] Lapczynski A, Bhatia SP, Letizia CS, Api AM. Fragrance material review on nerolidol (isomer unspecified). Food Chem Toxicol 2008, 46(Suppl 11), S247–250.

[30] Bartley JP, Foley P. Supercritical fluid extraction of Australian-grown ginger (*Zingiber officinale*). J Sci Food Agric 1994, 66(3), 365–371.

[31] Himmelbach A, Iten M, Grill E. Signalling of abscisic acid to regulate plant growth. Philos Trans R Soc Lond B Biol Sci 1998, 353(1374), 1439–1444.

[32] Feger W, Brandauer H, Ziegler H. Germacrenes in citrus peel oils. J Essent Oil Res 2001, 13(4), 274–277.

[33] Hanuš LO, Řezanka T, Dembitsky VM, Moussaieff A. Myrrh-*Commiphora* chemistry. Biomed Pap Med Fac Univ Palacky Olomouc Czech Repub 2005, 149(1), 3–28.

[34] Joshi SC, Padalia RC, Bisht DS, Mathela CS. Furanosesquiterpenoids from *Lindera pulcherrima* (Nees.) Benth. Ex Hook F Nat Prod Commun 2007, 2(9), 1934578X0700200914. 10.1177/1934578x0700200914.

[35] Srivastava AK, Srivastava SK, Syamsundar KV. Bud and leaf essential oil composition of *Syzygium aromaticum* from India and Madagascar. Flavour Fragr J 2005, 20(1), 51–53.

[36] Miyazawa M, Shimamura H, Nakamura S, Kameoka H. Antimutagenic activity of (+)-β-eudesmol and paeonol from *Dioscorea japonica*. J Agric Food Chem 1996, 44(7), 1647–1650.

[37] ElSohly HN, Croom JEM, El-Feraly FS, El-Sherei MM. A large-scale extraction technique of artemisinin from *Artemisia annua*. J Nat Prod 1990, 53(6), 1560–1564.

[38] Rodriguez Brasco MF, Seldes AM, Palermo JA. Paesslerins A and B: Novel tricyclic sesquiterpenoids from the soft coral *Alcyonium paessleri*. Org Lett 2001, 3(10), 1415–1417.

[39] Kittakoop P. Anticancer drugs and potential anticancer leads inspired by natural products. Stud Nat Prod Chem 2015, 44, 251–307.

[40] Thongtan J, Kittakoop P, Ruangrungsi N, Saenboonrueng J, Thebtaranonth Y. New antimycobacterial and antimalarial 8,9-secokaurane diterpenes from *Croton kongensis*. J Nat Prod 2003, 66(6), 868–870.

[41] Gurgel LA, Sidrim JJ, Martins DT, Cechinel Filho V, Rao VS. *In vitro* antifungal activity of dragon's blood from *Croton urucurana* against dermatophytes. J Ethnopharmacol 2005, 97(2), 409–412.

[42] Gutbrod K, Romer J, Dörmann P. Phytol metabolism in plants. Prog Lipid Res 2019, 74, 1–17.

[43] Santos CC, Salvadori MS, Mota VG, Costa LM, de Almeida AA, de Oliveira GA, Costa JP, de Sousa DP, de Freitas RM, de Almeida RN. Antinociceptive and antioxidant activities of phytol *in vivo* and *in vitro* models. Neurosci J 2013, 2013, 949452. 10.1155/2013/949452.

[44] Hasan CM, Khan S, Jabbar A, Rashid MA. Nasimaluns A and B: Neo-clerodane diterpenoids from *Barringtonia racemosa*. J Nat Prod 2000, 63(3), 410–411.

[45] Wijesekera K. A bioactive diterpene; Nasimalun A from *Croton oblongifolius* Roxb. Prayogik Rasayan 2017, 1(2), 41–44.

[46] Roengsumran S, Pornpakakul S, Muangsin N, Sangvanich P, Nhujak T, Singtothong P, Chaichit N, Puthong S, Petsom A. New halimane diterpenoids from *Croton oblongifolius*. Planta Med 2004, 70(1), 87–89.

[47] Roengsumran S, Petsom A, Sommit D, Vilaivan T. Labdane diterpenoids from *Croton oblongifolius*. Phytochemistry 1999, 50(3), 449–453.

[48] Dettrakul S, Kittakoop P, Isaka M, Nopichai S, Suyarnsestakorn C, Tanticharoen M, Thebtaranonth Y. Antimycobacterial pimarane diterpenes from the fungus *Diaporthe* sp. Bioorg Med Chem Lett 2003, 13(7), 1253–1255.

[49] Cambie RC, Mander LN. Chemistry of the Podocarpaceae – VI: Constituents of the heartwood of *Podocorpus totara* G. Benn Tetrahedron 1962, 18(4), 465–475.

[50] Brian PW, Elson GW, Hemming HG, Radley M. The plant-growth-promoting properties of gibberellic acid, a metabolic product of the fungus *Gibberella fujikuroi*. J Sci Food Agric 1954, 5(12), 602–612.

[51] Hayashi T, Kawasaki M, Miwa Y, Taga T, Morita N. Antiviral agents of plant origin. III.: Scopadulin, a novel tetracyclic diterpene from *Scoparia dulcis* L. Chem Pharm Bull 1990, 38(4), 945–947.

[52] Roengsumran S, Achayindee S, Petsom A, Pudhom K, Singtothong P, Surachetapan C, Vilaivan T. Two new cembranoids from *Croton oblongifolius*. J Nat Prod 1998, 61(5), 652–654.

[53] Roengsumran S, Singtothong P, Pudhom K, Ngamrochanavanich N, Petsom A, Chaichantipyuth C. Neocrotocembranal from *Croton oblongifolius*. J Nat Prod 1999, 62(8), 1163–1164.

[54] Inoue Y, Sakai M, Yao Q, Tanimoto Y, Toshima H, Hasegawa M. Identification of a novel casbane-type diterpene phytoalexin, ent-10-oxodepressin, from rice leaves. Biosci Biotechnol Biochem 2013, 77(4), 760–765.

[55] Wani MC, Taylor HL, Wall ME, Coggon P, McPhail AT. Plant antitumor agents. VI. Isolation and structure of taxol, a novel antileukemic and antitumor agent from *Taxus brevifolia*. J Am Chem Soc 1971, 93(9), 2325–2327.

[56] Chan PC, Xia Q, Fu PP. *Ginkgo biloba* leave extract: Biological, medicinal, and toxicological effects. J Environ Sci Health Part C- Environ Carcinog Ecotoxicol Rev 2007, 25(3), 211–244.

[57] Groweiss A, Look SA, Fenical W. Solenolides, new antiinflammatory and antiviral diterpenoids from a marine octocoral of the genus *Solenopodium*. J Org Chem 1988, 53(11), 2401–2406.

[58] De Giulio A, De Rosa S, Di Vincenzo G, Strazzullo G, Zavodnik N. Norsesterterpenes from the north Adriatic sponge *Ircinia oros*. J Nat Prod 1990, 53(6), 1503–1507.

[59] De Silva ED, Scheuer PJ. Manoalide, an antibiotic sesterterpenoid from the marine sponge *Luffariella variabilis* (Polejaeff). Tetrahedron Lett 1980, 21(17), 1611–1614.

[60] Rudi A, Yosief T, Schleyer M, Kashman Y. Bilosespens A and B: Two novel cytotoxic sesterpenes from the marine sponge *Dysidea cinerea*. Org Lett 1999, 1(3), 471–472.

[61] Buchanan MS, Edser A, King G, Whitmore J, Quinn RJ. Cheilanthane sesterterpenes, protein kinase inhibitors, from a marine sponge of the genus *Ircinia*. J Nat Prod 2001, 64(3), 300–303.

[62] Acebey L, Sauvain M, Beck S, Moulis C, Gimenez A, Jullian V. Bolivianine, a new sesterpene with an unusual skeleton from *Hedyosmum angustifolium*, and its isomer, isobolivianine. Org Lett 2007, 9(23), 4693–4696.

[63] Tsujimoto M. Squalene: A highly unsaturated hydrocarbon in shark liver oil. Ind Eng Chem 1920, 12(1), 63–72.

[64] Godtfredsen WO, Rastrup-Andersen N, Vangedal S, Ollis WD. Metabolites of *Fusidium coccineum*. Tetrahedron 1979, 35(20), 2419–2431.

[65] Dewick PM. Medicinal natural products: A biosynthetic approach. New York, USA. John Wiley & Sons. 2002.

[66] Risley JM. Cholesterol biosynthesis: Lanosterol to cholesterol. J Chem Educ 2002, 79(3), 377–384.

[67] Jayaprakasam B, Seeram NP, Nair MG. Anticancer and antiinflammatory activities of cucurbitacins from *Cucurbita andreana*. Cancer Lett 2003, 189(1), 11–16.

[68] Abdelwahab SI, Hassan LEA, Sirat HM, Yagi SM, Koko WS, Mohan S, Taha MM, Ahmad S, Chuen CS, Narrima P, Rais MM, Hadi AH. Anti-inflammatory activities of cucurbitacin E isolated from *Citrullus lanatus* var citroides: Role of reactive nitrogen species and cyclooxygenase enzyme inhibition. Fitoterapia 2011, 82(8), 1190–1197.

[69] Shao Y, Ho CT, Chin CK, Badmaev V, Ma W, Huang MT. Inhibitory activity of boswellic acids from *Boswellia serrata* against human leukemia HL-60 cells in culture. Planta Med 1998, 64(4), 328–331.

[70] Siddiqui MZ. *Boswellia serrata*, a potential antiinflammatory agent: An overview. Indian J Pharm Sci 2011, 73(3), 255–261.

[71] Aragão GF, Carneiro LMV, Junior APF, Vieira LC, Bandeira PN, Lemos TLG, Viana G S de B. A possible mechanism for anxiolytic and antidepressant effects of alpha-and beta-amyrin from *Protium heptaphyllum* (Aubl.) March. Pharmacol Biochem Behav 2006, 85(4), 827–834.

[72] Guo S, Kenne L, Lundgren LN, Rönnberg B, Sundquist BG. Triterpenoid saponins from *Quillaja saponaria*. Phytochemistry 1998, 48(1), 175–180.

[73] Safayhi H, Sailer E-R. Anti-inflammatory actions of pentacyclic triterpenes. Planta Med 1997, 63(6), 487–493.

[74] Kashiwada Y, Hashimoto F, Cosentino LM, Chen C-H, Garrett PE, Lee K-H. Betulinic acid and dihydrobetulinic acid derivatives as potent anti-HIV agents. J Med Chem 1996, 39(5), 1016–1017.

[75] Eiznhamer DA, Xu ZQ. Betulinic acid: A promising anticancer candidate. IDrugs Investig Drugs J 2004, 7(4), 359–373.

[76] Tapiero H, Townsend DM, Tew KD. The role of carotenoids in the prevention of human pathologies. Biomed Pharmacother 2004, 58(2), 100–110.

[77] Landrum JT, Bone RA. Lutein, zeaxanthin, and the macular pigment. Arch Biochem Biophys 2001, 385(1), 28–40.

[78] Valadon LRG, Mummery RS. Carotenoids of lilies and of red pepper: Biogenesis of capsanthin and capsorubin. Z Pflanzenphysiol 1977, 82(5), 407–416.

[79] Maoka T, Goto Y, Isobe K, Fujiwara Y, Hashimoto K, Mochida K. Antioxidative activity of capsorubin and related compounds from paprika (*Capsicum annuum*). J Oleo Sci 2001, 50(8), 663–665.

Kanchana Wijesekera, Sewwandi Subasinghe

5 Alkaloids

5.1 Introduction

Alkaloids are a group of natural products possessing one or more "nitrogen" atoms in the structure. Members of this group have contributed to medicine much more than any other group of natural products. Plants and fungi that contain alkaloids were used by many early communities around the world to alleviate pain or to achieve enhanced sensory experiences. However, it was the German pharmacist Wilhelm Meissner who first defined the term "alkaloids" in 1818 for the organic compounds with plant origin and had alkaline character [1]. Even though many alkaloids are of plant origin, several alkaloids are found in fungi, bacteria, marine organisms, amphibians, and even in humans as well [2–4]. Following compounds, (1)–(4) are obtained from a fungus: fly agaric (*Amanita muscaria*), a marine bacterium *Pseudovibrio denitrificans* Ab134, a marine tunicate *Ecteinascidia turbinata*, and an amphibian *Bufo alvarius* (Figure 5.1).

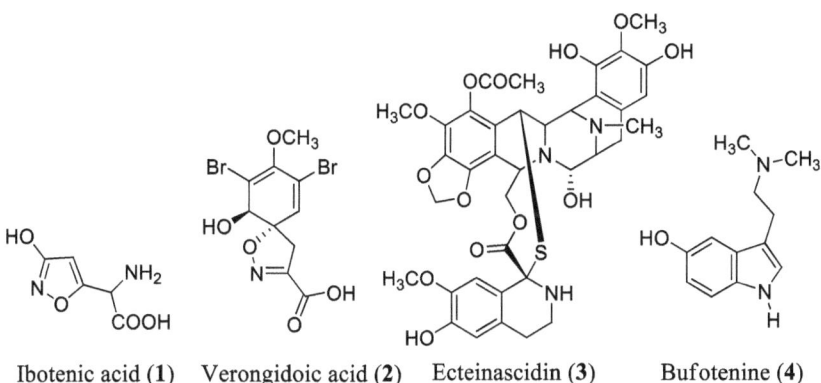

Ibotenic acid (**1**)   Verongidoic acid (**2**)   Ecteinascidin (**3**)   Bufotenine (**4**)

Figure 5.1: Some important alkaloids of natural origin.

Initial studies revealed that the nitrogen atom of the alkaloid is present in a heterocycle. Later with the advancement of the knowledge in this special group of compounds, alkaloids were defined as nitrogen-containing organic compounds with

Kanchana Wijesekera, Faculty of Allied Health Sciences, University of Ruhuna, Galle, Sri Lanka, e-mail: kdwijesekera@gmail.com
Sewwandi Subasinghe, Faculty of Allied Health Sciences,University of Ruhuna,Galle, Sri Lanka

https://doi.org/10.1515/9783110595949-005

complex molecular structures and significant biological activity. However, there are exceptions as in the case of ephedrine (**5**) (Figure 5.2).

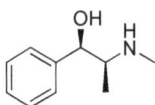

Ephedrine (**5**)

Figure 5.2: Ephedrine from *Ephedra sinica*.

Ephedrine obtained from *Ephedra sinica* is used to relieve asthma and it has a simple molecular structure and N is not present in a heterocycle. Due to its significant pharmacological activity, ephedrine is considered an alkaloidal amine.

Alkaloids are found in many plant families including Apocynaceae, Berberidaceae, Papaveraceae, etc. and isolated from various plant parts such as seeds (*Strychnos nux-vomica*), leaves (*Nicotiana tabacum*), tubers (*Gloriosa superba*), etc. Alkaloidal content of the plant varies with season, plant part, stage of the plant growth, and also during day and night [5–8].

Usually, alkaloids are present in plants as salts of organic acids such as oxalic acid, succinic acid, malic acid, etc. [9]. In addition, some alkaloids conjugated with sugars to form glycosides (α-tomatine) [10]. Morphine sulfate from *Papaver somniferum* is an example of an alkaloid salt with an inorganic acid.

Compared to other natural products, many alkaloids exhibit isomerism and two isomers even show different pharmacological activities (Figure 5.3). Quinine (**6**) and quinidine (**7**) are two enantiomers with different biological activities. Quinine (**6**) is an antimalarial drug and quinidine (**7**) is an antiarrhythmic agent.

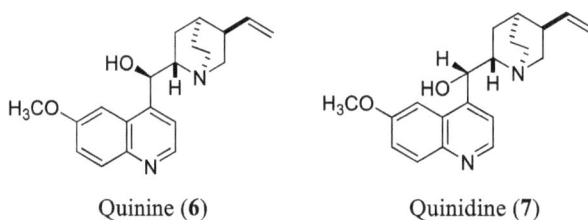

Quinine (**6**) Quinidine (**7**)

Figure 5.3: Structure of two enantiomers, quinine, and quinidine.

Alkaloids are biosynthesized from several different amino acids. Cyclization and modification of these amino acids together with other amino acids lead to diverse alkaloid structures (Table 5.1).

Table 5.1: Examples of some amino acids and their alkaloid derivatives.

Amino acid	Alkaloid derivative
L-Ornithine	
	(−)−Hygrine
L-Lysine	
	Lobeline
L-Tyrosine	
	Mescaline
L-Tryptophan	
	Psilocin
L-Histidine	
	Pilocarpine

5.2 Classification of alkaloids

Over 12,000 alkaloids have been identified in plants [9]. These alkaloids can be classified according to their preliminary point of biosynthesis, chemical structure, taxonomy, and pharmacological activity (Table 5.2). In this chapter, some important members of several different classes of alkaloids and their activities will be discussed.

Table 5.2: Classification of alkaloids.

Biosynthetic classification	Chemical classification
Example	Example
(i) Tyrosine derived alkaloids – Phenylethylamine	(i) Quinoline alkaloids – Quinine
(ii) Tryptophan derived alkaloids – Indole	(ii) Indole alkaloids – Vincristine
(iii) Ornithine derived alkaloids – Pyrrolidine	(iii) Tropane alkaloids – Cocaine
(iv) Lysine derived alkaloids – Piperidine	(iv) Isoquinoline alkaloids – Morphine
(v) Histidine derived alkaloids – Imidazole	(v) Alkaloidal amines – Ephedrine
Taxonomical classification	**Pharmacological classification**
Example	Example
(i) Cannabinaceous alkaloids – from *Cannabis sativa*	(i) Bronchodilator – Ephedrine
(ii) Rubiaceous alkaloids – from *Cinchona* sp.	(ii) Analgesic – Morphine
(iii) Solanaceous alkaloids – from *Atropa belladonna*	(iii) Anticancer – Vincristine
	(iv) Antimalarial – Quinine
	(v) Anti-inflammatory – Colchicine

5.2.1 Phenylalkylamine alkaloids

Compounds belonging to this group of alkaloids do not contain a nitrogen atom in a cycle. However, these compounds possess significant biological activity (Table 5.3).

Table 5.3: Phenylalkylamine alkaloids and their pharmacological activity.

Alkaloid	Pharmacological activity
Ephedrine ($C_{10}H_{15}NO$)	This alkaloid is isolated from the medicinal plant used in Chinese Medicine, Ma Huang (*Ephedra sinica*). The structure is similar to adrenaline, which is a natural hormone. Due to the bronchodilator properties, ephedrine is used to treat asthma.
Colchicine ($C_{22}H_{25}NO_6$)	This alkaloid is isolated from the meadow saffron (*Colchicum autumnale*) and several other species of *Colchicum*. Colchicine has a long history in medicine as a specific drug for the treatment of gout and rheumatism. Colchicine acts by binding to tubulin causing microtubule to depolymerize. As neutrophils lose their mobility, it helps to reduce inflammation and arthritis. It also possesses anti-tumor activity [11].

Table 5.3 (continued)

Alkaloid	Pharmacological activity
Mescaline ($C_{11}H_{17}NO_3$) H₃CO / H₃CO / OCH₃ / NH₂ structure	This is an alkaloid isolated from the heads of the peyote cactus *Lophophora williamsii* which causes hallucination. The cactus is used by the traditional population in Mexico to alleviate several illnesses [12–14].

5.2.2 Pyridine, piperidine, and pyrrolizidine alkaloids

Alkaloids that are derived from nicotinic acid which is biosynthesized via L-lysine are known as pyridine alkaloids. A well-known example of this family is the alkaloid nicotine isolated from *Nicotiana tabacum*.

The same amino acid L-lysine cyclizes and converts into different intermediates which ultimately leads to the formation of alkaloids having a piperidine nucleus. Lobeline and lobelanine from *Lobelia inflata*, are a few examples for this group (Table 5.4).

Pyrrolizidine alkaloids exhibit potent toxicity, and these alkaloids-containing plants are probably the most common poisonous plants affecting livestock and human.

Table 5.4: Pyridine, piperidine and pyrrolizidine alkaloids and their pharmacological activity.

Alkaloid	Pharmacological activity
Nicotine ($C_{10}H_{14}N_2$) nicotine structure	This is the most widely studied member of the pyridine group of alkaloids. It is isolated from tobacco (*Nicotiana tabacum*). It possesses neuronal stimulant properties and long-term use causes addiction. It is included in several tobacco preparations such as cigarettes. Nicotine is formulated into chewing gum as an aid for smoking cessation [15]. Nornicotine and anabasine are two other examples of nicotine-type alkaloids isolated from tobacco.

Table 5.4 (continued)

Alkaloid	Pharmacological activity
Arecoline ($C_8H_{13}NO_2$)	Piperidine-type alkaloid arecoline is one of the important alkaloids isolated from betel nuts (*Areca catechu*), a plant that grows in countries like Sri Lanka, India, and Malaysia. These betel nuts are consumed by farmers to alleviate fatigue as arecoline has a stimulant effect on the central nervous system (CNS). Other alkaloids that share a similar structure to arecoline include arecaidine, guvacine, and guvacoline. Arecoline is used in veterinary practice as a vermicide [16].
Coniine ($C_8H_{17}N$)	This is a poisonous piperidine alkaloid isolated from hemlock (*Conium maculatum*). Symptoms of acute poisoning include muscular weakness, pupil dilation, excess salivation, etc. Humans can be exposed to coniine via the food chain [17]. The plant is famous in history as the great Greek philosopher Socrates was forced to drink the preparation of hemlock for his execution [18].
Lobeline ($C_{22}H_{27}NO_2$)	Lobeline is an alkaloid isolated from the leaves and tops of wild tobacco or puke weed (*Lobelia inflata*). As lobeline exerts effects similar to nicotine, it is involved in the cessation of smoking [19].
Piperine ($C_{17}H_{19}NO_3$)	Piperine is one of the important alkaloids that belong to the piperidine family. It is isolated from the common spice black pepper (*Piper nigrum*). It exhibits many pharmacological activities including antidiarrheal, antidepressant, antihypertensive effect, antitumor activity, etc. [20].
Pipernonaline ($C_{21}H_{27}NO_3$)	This is a piperidine-type alkaloid obtained from the dried fruits of *Piper longum*. Pipernonaline exhibited insecticidal activity against *Spodoptera litura* [21]. In addition, it showed larvicidal activity against *Culex pipiens* [21].

Table 5.4 (continued)

Alkaloid	Pharmacological activity
(-)-Spectaline ($C_{20}H_{39}NO_2$)	This is a cytotoxic piperidine-type alkaloid isolated by bioassay-guided fractionation of a bioactive extract of the Brazilian legume, *Cassia leptophylla* [22].
Senecionine ($C_{18}H_{25}NO_5$)	This pyrrolizidine alkaloid isolated from *Senecio vulgaris* exhibited toxicity via causing hepatic necrosis *in vivo*. Moreover, this alkaloid can bind with deoxyguanosine to form DNA adducts [23, 24].
Castanospermine ($C_8H_{15}NO_4$)	This is an important alkaloid containing a piperidine ring system isolated from the seeds of the Australian legume, *Castanospermum australe* Castanospermine is a potent inhibitor of fibroblast α and β glucosidases, and found to be effective against HIV [25, 26].

5.2.3 Quinoline alkaloids

Alkaloids belonging to this family are characterized by the presence of a quinoline nucleus in the molecule. The most remarkable member of this family is the antimalarial drug quinine from the bark of the cinchona tree, and it provided the pharmacophore for the synthesis of other antimalarial drugs. Usually, quinoline alkaloids are derived from anthranilic acid except for cinchona alkaloids and camptothecin which are derived from tryptophan (Table 5.5).

Table 5.5: Quinoline alkaloids and their pharmacological activity.

Alkaloid	Pharmacological activity
Quinine and Quinidine ($C_{20}H_{24}N_2O_2$) Quinine Quinidine	Both these alkaloids are enantiomers and were isolated from the barks of *Cinchona* sp. Quinine as a potent antimalarial drug provided the pharmacophore for the synthesis of other antimalarial drugs such as quinacrine, chloroquine, and mefloquine. Quinidine is employed in the treatment of cardiac arrhythmias [27].

Table 5.5 (continued)

Alkaloid	Pharmacological activity
Camptothecin ($C_{20}H_{16}N_2O_4$)	This is a pentacyclic, antitumor alkaloid isolated from *Camptotheca acuminata*, a tree native to China. Due to very low solubility, high toxicity, and rapid inactivation, the therapeutic action of this alkaloid was poor. However, camptothecin provided the pharmacophore for the synthesis of more water-soluble alkaloids topotecan and irinotecan which were encountered as chemotherapeutic agents [28].
Galipine ($C_{20}H_{21}NO_3$)	Galipine is a quinolone-type alkaloid isolated from the bark of a shrub that grows in tropical America *Galipea officinalis* (Rutaceae). The bark is reputed in folk medicine as being antispasmodic, antipyretic, astringent, and tonic [29].

5.2.4 Isoquinoline alkaloids

The well-known example of the isoquinoline group of alkaloids is morphine, possibly the oldest narcotic known. Isoquinoline alkaloids are a diverse group of alkaloids and are de-rived from L-tyrosine. There are several alkaloidal subclasses which include simple tetrahydroisoquinoline, benzyltetrahydroisoquinoline, phenethylisoquinoline, etc (Table 5.6).

Table 5.6: Isoquinoline alkaloids and their pharmacological activity.

Alkaloid	Pharmacological activity
Morphine ($C_{17}H_{19}NO_3$) Morphine: $R_1 = R_2 = H$ Codeine: $R_1 = CH_3$, $R_2 = H$	This is one of the major alkaloids isolated from the dry latex of opium poppy (*Papaver somniferum*). Morphine is named after the Greek god Morpheus, the creator of sleep and dreams. In Western medicine, it is used as a potent narcotic analgesic. However, morphine is easily converted into other drugs of abuse such as heroin [17]. Codeine ($C_{18}H_{21}NO_3$) is another alkaloid that was isolated from the same plant, and both alkaloids share the same chemical structure with an extra methyl group in codeine. Pharmaceutically, codeine is employed in cough preparations.

Table 5.6 (continued)

Alkaloid	Pharmacological activity
Papaverine ($C_{20}H_{21}NO_4$)	This is another alkaloid isolated from opium poppy with an isoquinoline skeleton. Papaverine does not show addictive properties and it is used to treat spasms [30].
Emetine ($C_{29}H_{40}N_2O_4$) Emetine: R = CH_3 Cephaeline: R = H	Emetine is an isoquinoline type isolated from the dried roots of ipecac *Cephelis ipecacuanha*. Natives of Brazil used the roots of this plant to treat diarrhea. Cephaeline ($C_{28}H_{42}N_2O_3$) is another alkaloid that is isolated from the ipecac plant. Ipecac has been used as an emetic, an amoebicide, and an expectorant [31]. Emetine has a more expectorant and less emetic action than cephaëline [32].
Tubocurarine ($C_{37}H_{41}N_2O_6$)	The isoquinoline-type alkaloid tubocurarine is obtained from the climbing plant *Chondrodendron tomentosum* and was used as an arrowhead poison (curare) by South American Indians. It acts as a neuromuscular blocker and it has been suggested that these alkaloids function as herbivore deterrents [33]. This compound was also the template for other muscle relaxants like atracurium [34].
Berberine ($C_{20}H_{18}NO_4$)	This is an alkaloid found in many members of the families of Berberidaceae, Ranunculaceae, and other families. Plants that contain berberine have been used in traditional medicine throughout the world [33]. Berberine is reported to exhibit many biological activities, including antiamoebic, antibacterial, and anti-inflammatory properties. In addition, berberine is effective against osteosarcoma, lung, liver, prostate, and breast cancer.

Table 5.6 (continued)

Alkaloid	Pharmacological activity
Kreysigine (C$_{22}$H$_{27}$NO$_5$)	This is an alkaloid isolated from several genera including *Androcymbium*, *Colchicum*, and *Kreysigia*. Both kreysigine and colchicine are biosynthesized from the same intermediate, autumnaline [35]
Hydrastine (C$_{21}$H$_{21}$NO$_6$)	This is an alkaloid isolated from *Hydrastis canadensis* which shows anti-inflammatory and hemostatic activities. [36]

5.2.5 Indole alkaloids

Indole alkaloids derived from L-tryptophan contain an indole ring system and are considered as a very important source of biologically active compounds (Table 5.7). The group consists of several subclasses which include simple indole alkaloids, terpenoid indole alkaloids, simple β-carboline alkaloids, etc. Many psychoactive compounds are structurally related to indole alkaloids. For example, psilocin (**8**) from the mushroom *Psilocybe mexicana* and lysergic acid diethylamide (LSD) (**9**) which is semi-synthesized from ergometrine (fungus *Claviceps purpurea*) (Figure 5.4). One of the pharmacologically significant alkaloids that belong to this group is the anti-cancer alkaloid vincristine.

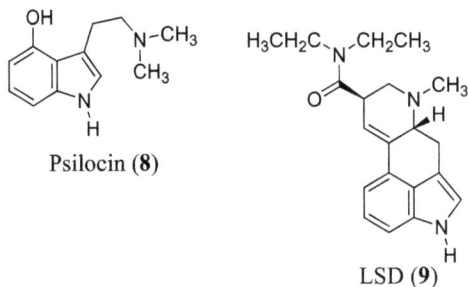

Psilocin (**8**)

LSD (**9**)

Figure 5.4: Two psychoactive alkaloids psilocin and LSD with indole moiety.

Table 5.7: Indole alkaloids and their pharmacological activity.

Alkaloid	Pharmacological activity
Strychnine ($C_{21}H_{22}N_2O_2$) Strychnine: $R_1 = R_2 = H$ Brucine: $R_1 = R_2 = CH_3O$	Strychnine and brucine are indole-type alkaloids isolated from the seeds of nux-vomica (*Strychnos nux-vomica*) found in Sri Lanka and India. Both these compounds are highly poisonous and in the past, poisoning incidents were reported occasionally. In particular, strychnine has been used for the extermination of moles [32].
Reserpine ($C_{33}H_{40}N_2O_9$)	Ancient people in India and Sri Lanka have used Indian snakeroot (*Rauwolfia serpentaria*) for centuries as an antidote to poisonous snake bites and to treat madness. One of the major alkaloids isolated from this plant is reserpine. This alkaloid is clinically important in the treatment of schizophrenia. Ajmaline is another clinically important indole-type alkaloid isolated from the same plant [37].
Vincristine ($C_{46}H_{56}N_4O_{10}$) Vincristine: R = CHO Vinblastine: R = CH_3	The most important alkaloids belonging to the indole group of alkaloids are the anticancer agents vincristine ($C_{46}H_{56}N_4O_{10}$) and vinblastine ($C_{46}H_{58}N_4O_9$) from the Madagascar periwinkle (*Catharanthus roseus*). These are complex dimeric indoles that exert anticancer activity by inhibiting the polymerization of tubulin [38]. Vinblastine also provided the pharmacophore for the synthesis of another effective anticancer drug vindesine [39].
Yohimbine ($C_{21}H_{26}N_2O_3$)	Yohimbine is an alkaloid isolated from the African plant *Pausinystalia johimbe* and acts as an α-adrenergic blocker. Further, it has a wide reputation as a sexual stimulant [40, 41].

Table 5.7 (continued)

Alkaloid	Pharmacological activity
Harmine ($C_{13}H_{12}N_2O$) 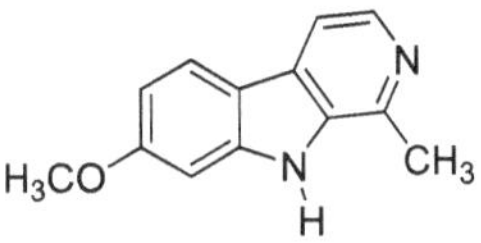 	Harmine is an indole-type alkaloid with a β-carboline skeleton isolated from the Syrian Rue (*Peganum harmala*). This alkaloid showed potent cytotoxicity against several cancer cell lines which include KB, A549, CAKI-I, 1A9, and HEL cells [42, 43].
Ibogaine ($C_{20}H_{26}N_2O$) 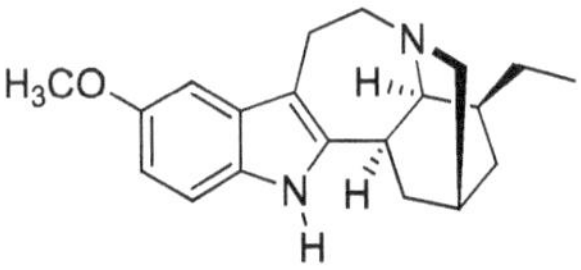 	This is an indole-type alkaloid isolated from the plant iboga (*Tabernanthe iboga*). Ibogaine is known to act on the central nervous system with hallucination and exhibit an amphetamine-like effect. Ibogaine consists of a spectrum of anti-addictive properties with opiate, cocaine, and alcohol [44].
Physostigmine ($C_{15}H_{21}N_3O_2$) 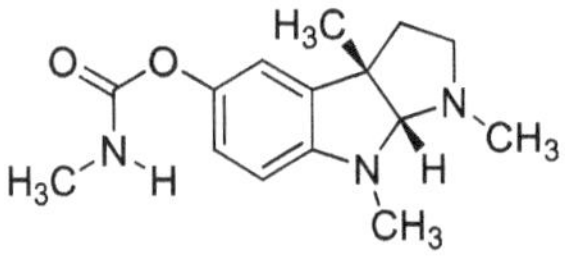 	This is a toxic indole alkaloid isolated from the African calabar bean (*Physostigma venenosum*). The toxic component of this species is physostigmine where it acts as an inhibitor of acetylcholinesterase. This alkaloid has a special interest in the treatment of Alzheimer's disease [45, 46]

5.2.6 Tropane alkaloids

Tropane alkaloids are bicyclic compounds biosynthesized by L-ornithine and known to originate in several plant families including Solanaceae and Erythroxylaceae (Figure 5.5). Several tropane-type alkaloids are used in medicine. However, many of them are known for their toxicity [17].

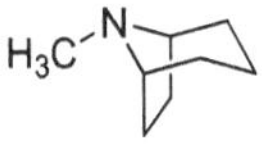

Figure 5.5: Tropane moiety.

The contemporary pharmaceutical industry manufactures over 20 active pharmaceutical substances containing tropane moiety in their structure, which are applied as mydriatics, antiemetics, antispasmodics, anesthetics, and bronchodilators [47] (Table 5.8).

Table 5.8: Tropane alkaloids and their pharmacological activity.

Alkaloid	Pharmacological activity
Atropine ($C_{17}H_{23}NO_3$)	The plant deadly nightshade (*Atropa belladonna*) produces hyoscyamine which occurs in the plant as a racemic mixture. This mixture is usually referred to as atropine. In addition to *Atropa* sp., this alkaloid is also isolated from several species like *Datura*, henbane (*Hyoscyamus niger*), and *Duboisia* sp. [48]. Atropine act as an anticholinergic drug and also as a mydriatic, to dilate the pupils [49].
Hyoscine ($C_{17}H_{21}NO_4$)	Hyoscine or (-)-scopolamine is the epoxide derivative of hyoscyamine isolated from the same species as atropine. Hyoscine is a useful prophylaxis against nausea and vomiting after middle ear surgery [50].
Cocaine ($C_{17}H_{21}NO_4$)	This is one of the tropane-type alkaloids isolated from the coca plant (*Erythroxylum coca*) which is grown in several South American countries. It is reported that Peruvian Indians used coca for at least 1,000 years before the arrival of Europeans. Cocaine is used as a local anesthetic in ophthalmology, a central nervous system stimulant and to improve physical strength [17]. However, it is a drug of abuse.
Calystegine B_2 ($C_7H_{13}NO_4$)	Calystegines are a group of nor-tropane alkaloids (lack of carbon on nitrogen) isolated from several plant families including Solanaceae (potato – *Solanum tuberosum*) and Convolvulaceae (sweet potato – *Ipomoea batatas*). This group includes calystegine A_3, A_5, B_1, B_2, B_3, B_4, C_1, etc. [51]. It was reported that calystegine B_2 is a potent competitive inhibitor of glucosidase [52].

5.2.7 Purine alkaloids

Most of the nonalcoholic beverages that we consume daily contain alkaloids belonging to the purine family. Caffeine (**10**) is an important member of the group and is available in tea (*Camellia sinensis*) and coffee (*Coffea arabica*). Theophylline (**11**) is another purine alkaloid that also occurs in minor amounts in tea. Caffeine increases alertness and reduces fatigue [53]. Theobromine (**12**) is also a commercially important purine alkaloid found in the seeds of cacao (*Theobroma cacao*) (Figure 5.6).

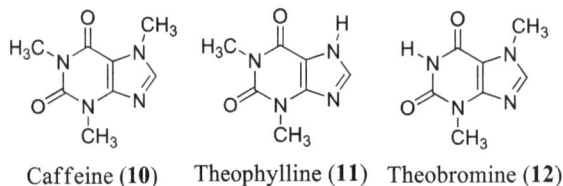

Caffeine (**10**)   Theophylline (**11**)   Theobromine (**12**)

Figure 5.6: Some important purine alkaloids.

5.2.8 Steroidal alkaloids

Several steroidal alkaloids are found in the plant kingdom including in the families of Solanaceae, Liliaceae, and Apocynaceae and are known to have different biological activities. Alkaloids belonging to this family are biosynthesized via the insertion of one or two N atoms into a steroid molecule. They can be classified broadly into solanum alkaloids and veratrum alkaloids.

Veratrum alkaloids are biosynthesized by the expansion of ring D in the steroidal nucleus at the expense of ring C which ultimately becomes five-membered (Figure 5.7) [33].

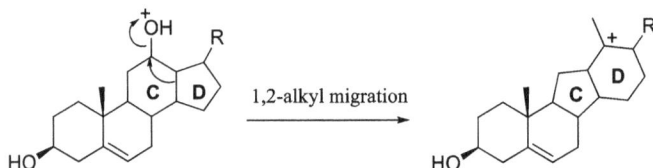

Figure 5.7: Formation of veratrum alkaloids.

Examples of solanum alkaloids are solasodine from *Capsicum annuum*, tomatidine from *Lycopersicon esculentum*, and solanidine from *Solanum tuberosum*; all of which occurs in the plants as glycosides. Kurchi bark (*Holarrhena pubescens*) has been valued for its antidysenteric properties for a long time and is responsible for the production of the steroidal alkaloid conessine [54].

Two distinct chemical groups of veratrum alkaloids can be identified and these are now referred as the jerveratrum and ceveratrum groups. Jerveratrum alkaloids contain only 1–3 oxygen atoms whereas ceveratrum alkaloids are highly hydroxylated compounds with 7–9 oxygen atoms. Jervine and cyclopamine are examples of toxic jerveratrum alkaloids and isolated from *Veratrum califomicum*. Protoveratrine A and protoveratrine B are examples of ceveratrum alkaloids (Figure 5.8).

Steroidal alkaloids and their glycosides are reported to have varieties of bioactivities including antimicrobial, anti-inflammatory, antinociceptive, etc. [32, 55, 56].

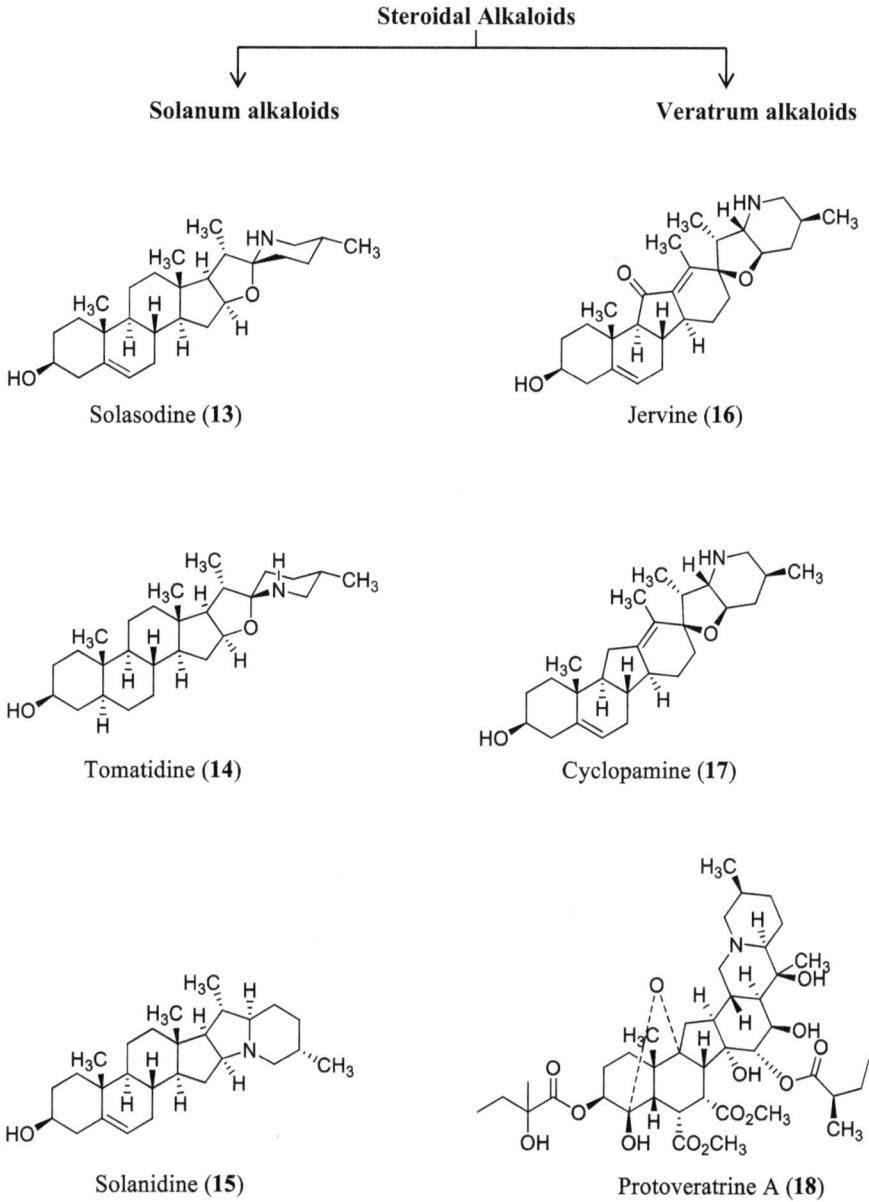

Figure 5.8: Some important steroidal alkaloids.

5.3 Summary

Natural products, especially alkaloids, have been playing a vital role in the prevention and treatment of many diseases in the world. Opium is possibly the oldest narcotic known, for as early as 4000 BC. Alkaloids-containing plants, fungi, etc. have been used by native people in early communities of the world to relieve pain, to gain physiological satisfaction, and also used for religious ceremonies. Structures of many alkaloids were not known for centuries. With the industrial revolution, there was a rapid development in the isolation and characterization of the active compounds of the crude drugs which were already in use. Morphine was the first alkaloid to isolate in the crystalline form by Friedrich Sertürner in 1804. Since then, a huge number of alkaloids have been isolated from many natural resources and plant pioneered in alkaloid biosynthesis. Due to the presence of nitrogen in the molecule, alkaloids exhibit many pharmacological activities such as antiarrhythmic, antimalarial, antihypertensive, antitumor, analgesic, and antiprotozoal. In many instances, alkaloids themselves act as drugs or sometimes, they provide pharmacophores for the synthesis of new drugs. Therefore, exploring natural products for their potent biological activities will remain a crucial part of the drug discovery process.

References

[1] Jones AW. Early drug discovery and the rise of pharmaceutical chemistry. Drug Test Anal 2011, 3(6), 337–344.

[2] Nicacio KJ, Ióca LP, Fróes AM. et al., Cultures of the marine bacterium *Pseudovibrio denitrificans* Ab134 produce bromotyrosine-derived alkaloids previously only isolated from marine sponges. J Nat Prod 2017, 80(2), 235–240.

[3] Schwartsmann G, Da Rocha AB, Berlinck RG, Jimeno J. Marine organisms as a source of new anticancer agents. Lancet Oncol 2001, 2(4), 221–225.

[4] McBride MC. Bufotenine: Toward an understanding of possible psychoactive mechanisms. J Psychoactive Drugs 2000, 32(3), 321–331.

[5] Pěnčíková K, Urbanová J, Musil P, Táborská E, Gregorová J. Seasonal variation of bioactive alkaloid contents in *Macleaya microcarpa* (Maxim.) Fedde. Molecules 2011, 16(4), 3391–3401.

[6] Johnson EL, Emche SD. Variation of alkaloid content in *Erythroxylum coca* leaves from leaf bud to leaf drop. Ann Bot 1994, 73(6), 645–650.

[7] Verma V, Kasera PK. Variations in secondary metabolites in some arid zone medicinal plants in relation to season and plant growth. Indian J Plant Physiol 2007, 12(2), 203–206.

[8] Itenov K, Mølgaard P, Nyman U. Diurnal fluctuations of the alkaloid concentration in latex of poppy *Papaver somniferum* is due to day–night fluctuations of the latex water content. Phytochemistry 1999, 52(7), 1229–1234.

[9] Rolin D. Metabolomics coming of age with its technological diversity. London, UK, Academic Press, 2012.

[10] Roddick JG. The steroidal glycoalkaloid α-tomatine. Phytochemistry 1974, 13(1), 9–25.

[11] Alali FQ, El-Alali A, Tawaha K, El-Elimat T. Seasonal variation of colchicine content in *Colchicum brachyphyllum* and *Colchicum tunicatum* (Colchicaceae). Nat Prod Res 2006, 20(12), 1121–1128.

[12] Aizpurua-Olaizola O, Soydaner U, Öztürk E, Schibano D, Simsir Y, Navarro P, Etxebarria N, Usobiaga A. Evolution of the cannabinoid and terpene content during the growth of *Cannabis sativa* plants from different chemotypes. J Nat Prod 2016, 79(2), 324–331.

[13] Cunningham N. Hallucinogenic plants of abuse. Emerg Med Australas 2008, 20(2), 167–174.

[14] Schultes RE. The appeal of peyote (*Lophophora williamsii*) as a medicine. Am Anthropol 1938, 40(4), 698–715.

[15] Jarvis MJ, Raw M, Russell MA, Feyerabend C. Randomised controlled trial of nicotine chewing-gum. Br Med J (Clin Res Ed) 1982, 285(6341), 537–540.

[16] Bahmani M, Rafieian-Kopaei M, Hassanzadazar H, Saki K, Karamati SA, Delfan B. A review on most important herbal and synthetic antihelmintic drugs. Asian Pac J Trop Med 2014, 7, S29–33.

[17] Crozier A, Clifford MN, Ashihara H. Plant Secondary Metabolites: Occurrence, Structure And Role In The Human Diet. New Jersey, USA, John Wiley & Sons, 2006.

[18] Arihan O, Arihan SK, Touwaide A. The Case Against Socrates and His Execution. Wexler P, ed, History of Toxicology and Environmental Health -Toxicology in Antiquity 1st ed. Amsterdam, NL, Elsevier, 2014, 69–82.

[19] Heinrich M, Barnes J, Prieto-Garcia J, Gibbons S, Williamson EM. Fundamentals of Pharmacognosy and Phytotherapy. Amsterdam, NL, Elsevier, 2017.

[20] Singh A, Duggal S. Piperine-Review of advances in pharmacology. Int J Pharm Sci Nanotechnol 2009, 2(3), 615–620.

[21] Park B-S, Lee S-E, Choi W-S, Jeong C-Y, Song C, Cho K-Y. Insecticidal and acaricidal activity of pipernonaline and piperoctadecalidine derived from dried fruits of *Piper longum* L. Crop Prot 2002, 21(3), 249–251.

[22] Bolzani VDS, Gunatilaka AAL, Kingston DGI. Bioactive and other piperidine alkaloids from *Cassia leptophylla*. Tetrahedron 1995, 51(21), 5929–5934.

[23] Tu Z-B, Konno C, Soejarto DD, Waller DP, Bingel AS, Molyneux RJ, Edgar JA, Cordell GA, Fong HHS. Identification of senecionine and senecionine N-oxide as antifertility constituents in *Senecio vulgaris*. J Pharm Sci 1988, 77(5), 461–463.

[24] Fu PP, Xia Q, Lin G, Chou MW. Pyrrolizidine alkaloids – genotoxicity, metabolism enzymes, metabolic activation, and mechanisms. Drug Metab Rev 2004, 36(1), 1–55.

[25] Stevens KL, Molyneux RJ. Castanospermine – a plant growth regulator. J Chem Ecol 1988, 14(6), 1467–1473.

[26] Roja G, Heble MR. Castanospermine, an HIV inhibitor from tissue cultures of *Castanospermum australe*. Phyther Res 1995, 9(7), 540–542.

[27] Verpoorte R, Schripsema J, Van der Leer T. Cinchona alkaloids. Brossi A, ed, The alkaloids: Chemistry and Pharmacology-Vol 34. 1st New York, USA, Academic Press, 1988, 331–398.

[28] Li Q-Y, Zu Y-G, Shi R-Z, Yao L-P. Review camptothecin: Current perspectives. Curr Med Chem 2006, 13(17), 2021–2039.

[29] Rakotoson JH, Fabre N, Jacquemond-Collet I, Hannedouche S, Fourasté I, Moulis C. Alkaloids from *Galipea officinalis*. Planta Med 1998, 64(08), 762–763.

[30] Osman F, Buller N, Steeds R. Use of intra-arterial papaverine for severe arterial spasm during radial cardiac catheterization. J Invasive Cardiol 2008, 20(10), 551–552.

[31] Teshima D, Ikeda K, Shimomura K, Aoyama T. Simultaneous determination of emetine and cephaeline in ipecac syrup. Chem Pharm Bull 1989, 37(1), 197–199.

[32] Evans WC. Trease and Evans Pharmacognosy. Philadelphia, USA, Saunders Ltd, 2009.

[33] Dewick PM. Medicinal Natural Products: A Biosynthetic Approach. New Jersey, USA, John Wiley & Sons, 2002.

[34] Harvey AL, Bradley KN, Cochran SA, Rowan EG, Pratt JA, Quillfeldt JA, Jerusalinsky DA. What can toxins tell us for drug discovery?. Toxicon 1998, 36(11), 1635–1640.

[35] Tojo E. The homoaporphine alkaloids. J Nat Prod 1989, 52(5), 909–921.

[36] Anushri M, Yashoda R, Puranik MP. Herbs: A good alternatives to current treatments for oral health problems. Int J Adv Health Sci 2015, 1(12), 26–32.

[37] López-Muñoz F, Bhatara VS, Alamo C, Cuenca E. Historical approach to reserpine discovery and its introduction in psychiatry. Actas Esp Psiquiatr 2004, 32(6), 387–395.

[38] Duflos A, Kruczynski A, Barret JM. Novel aspects of natural and modified vinca alkaloids. Curr Med Chem Agents 2002, 2(1), 55–70.

[39] Vats T, Buchanan G, Mehta P, Ragab A, Hvizdale E, Nitschke R, Link M, Beardsley GP, Maybee D, Krischer J. A study of toxicity and comparative therapeutic efficacy of vindesine-prednisone vs. vincristine-prednisone in children with acute lymphoblastic leukemia in relapse, A Pediatric Oncology Group study. Invest New Drugs 1992, 10(3), 231–234.

[40] Ostojic SM. Yohimbine: The effects on body composition and exercise performance in soccer players. Res Sport Med 2006, 14(4), 289–299.

[41] Tam SW, Worcel M, Wyllie M. Yohimbine: A clinical review. Pharmacol Ther 2001, 91(3), 215–243.

[42] Kartal M, Altun ML, Kurucu S. HPLC method for the analysis of harmol, harmalol, harmine and harmaline in the seeds of *Peganum harmala* L. J Pharm Biomed Anal 2003, 31(2), 263–269.

[43] Ishida J, Wang HK, Bastow KF, Hu CQ, Lee KH. Antitumor agents 201. Cytotoxicity of harmine and β-carboline analogs. Bioorg Med Chem Lett 1999, 9(23), 3319–3324.

[44] Alper KR, Lotsof HS, Frenken GM, Luciano DJ, Bastiaans J. Treatment of acute opioid withdrawal with ibogaine. Am J Addict 1999, 8(3), 234–242.

[45] Eadie GS. The inhibition of cholinesterase by physostigmine and prostigmine. J Biol Chem 1942, 146, 85–93.

[46] Mohs RC, Davis BM, Johns CA, Mathé AA, Greenwald BS, Horvath TB, Davis KL. Oral physostigmine treatment of patients with Alzheimer's disease. Am J Psychiatry 1985, 142(1), 28–33.

[47] Grynkiewicz G, Gadzikowska M. Tropane alkaloids as medicinally useful natural products and their synthetic derivatives as new drugs. Pharmacol Rep 2008, 60(4), 439–463.

[48] Jakabová S, Vincze L, Á F, Kilár F, Boros B, Felinger A. Determination of tropane alkaloids atropine and scopolamine by liquid chromatography–mass spectrometry in plant organs of *Datura* species. J Chromatogr A 2012, 1232, 295–301.

[49] Behçet A. The source-synthesis-history and use of atropine. Eurasian J Emerg Med 2014, 13(1), 2–3.

[50] Honkavaara P. Effect of transdermal hyoscine on nausea and vomiting during and after middle ear surgery under local anaesthesia. Br J Anaesth 1996, 76(1), 49–53.

[51] Bekkouche K, Daali Y, Cherkaoui S, Veuthey JL, Christen P. Calystegine distribution in some solanaceous species. Phytochemistry 2001, 58(3), 455–462.

[52] Asano N, Kato A, Miyauchi M, Kizu H, Tomimori T, Matsui K, Nash RJ, Molyneux RJ. Specific α-galactosidase inhibitors, N-methylcalystegines- structure/activity relationships of calystegines from *Lycium chinense*. Eur J Biochem 1997, 248(2), 296–303.

[53] Nawrot P, Jordan S, Eastwood J, Rotstein J, Hugenholtz A, Feeley M. Effects of caffeine on human health. Food Addit Contam 2003, 20(1), 1–30.

[54] Houghton PJ, Dias Diogo ML. The conessine content of *Holarrhena pubescens* from Malawi. Int J Pharmacogn 1996, 34(4), 305–307.

[55] Jiang QW, Chen MW, Cheng KJ, Yu PZ, Wei X, Shi Z. Therapeutic potential of steroidal alkaloids in cancer and other diseases. Med Res Rev 2016, 36(1), 119–143.

[56] Benforado JM, Flacke W, Mosimann W, Swaine CR. Studies on veratrum alkaloids. XXIX. the action of some germine esters and of veratridine upon blood pressure, heart rate and femoral blood flow in the dog. J Pharmacol Exp Ther 1960, 130(3), 311–320.

K. G. N. P. Piyasena, M. M. Qader

6 Saponins

6.1 Introduction

Saponins are generally known as polar, non-volatile, surface-active bioorganic natural compounds that are widely distributed in nature, occurring primarily in plants (more than 500 plant genera), lower marine organisms (starfish, sea cucumber, sponges), invertebrates (mollusks), and bacteria [1–3]. The name "saponin" is derived from the Latin word *sapo*, which means "soap," because saponin molecules form soap-like detergent properties when shaken with water even at low concentrations [4, 5]. Therefore, plant materials, e.g. soapwort (*Saponaria officinalis*), quillaia or soapbark (*Quillaja saponaria*), soaproot (*Chlorogalum pomeridianum*), soapberry (*Sapindus saponaria*), and soap nut (*Sapindus mukorossi*) containing saponins, were historically used for the cleansing process. The amphoteric nature that has polar and non-polar moiety enhances the foaming, detergent, and emulsifying natures of saponins [3]. The compounds have the ability to lyse the red blood cells by increasing plasma membrane permeability; thus, they are toxic when contacted with bloodstream. But not all saponins are harmful. Some of our food and beverages, e.g. beans, lentils, soybeans, spinach, oats, fenugreek, ginseng, and tea, contain a significant amount of saponins, and they are nutritionally important. Saponins are biosynthesized in different parts of the plant, and their distribution of concentrations varies. Moreover, the saponin content in young plants is higher than the mature plants. Also depending on the biotic (cultivar, physiology) and abiotic (environmental, geographical) conditions the saponin content varies in the plant species [6]. As an example, in the garden marigold (family Asteraceae) contains about 3.5% of C_3-glucuronic acid saponins in flowers and C_3-glucose saponins about 2.5% in the roots [7].

Saponins consist of sugar moieties linked glycosidically (C–O-sugar bond) to the hydroxyl group at C-3 to a hydrophobic aglycone nucleus (Figure 6.1). Therefore, saponins are high molecular weight and high polar molecules. Hydrolysis of a saponin molecule yielded glycone (sugar moiety) and aglycone (non-sugar moiety). The diversity of the saponins is mainly due to the sugar moieties and the sapogenins present. The presence of carboxyl groups in the aglycone and the glycone moieties of the saponins make them acidic or neutral saponins.

K. G. N. P. Piyasena, Tea Research Institute of Sri Lanka, Talawakelle, Sri Lanka,
e-mail: nelumpriya@yahoo.com
M. M. Qader, Department of Chemistry, The Open University of Sri Lanka, Nawala, Nugegoda,
Sri Lanka; Present Address: Institute for Tuberculosis Research, University of Illinois at Chicago,
IL, United States

https://doi.org/10.1515/9783110595949-006

Sapogenin ---- O ---- Sugar

Aglycone unit ⸱⸱⸱⸱ Glycone unit
-hydrophobic nature - hydrophilic nature

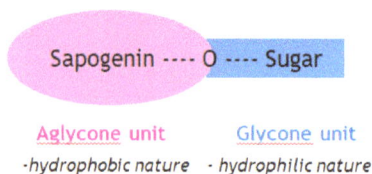

Figure 6.1: Saponin molecule.

6.1.1 Glycone: sugar unit

The common sugar moieties found in plants include L-arabinose (Ara), D-fucose (Fuc), D-glucose (Glc), D-galactose (Gal), L-rhamnose (Rha), D-glucuronic acid (GlucA), D-galacturonic acid (GalA), and D-xylose (Xyl). D-Quinovose (Qui), Glc, Ara, GlucA, and Xyl are the most common sugar moieties in saponins isolated from marine organisms. Glucose, arabinose, glucuronic acid, and xylose are most frequently directly attached glycone units to the sapogenin nucleus. The sugar units in the sapogenins could be linear or branched. The number of sugar units attached to the sapogenin unit can be varying up to eleven and it is the highest number of sugar units found naturally [8], with two to five sugar units being the most abundantly found in nature. Furthermore, glycosidic linkages to the sapogenin moiety through α- and β-hydroxy groups, presence of pentose and hexose sugars, and d- (dextro) and l- (leavo) isomers of sugars can result in a very diverse group of compounds.

The saponins are categorized into three major classes according to the structure of aglycone unit: triterpenoid glycosides, steroidal glycosides, and alkaloid steroid glycosides [9, 10].

6.1.2 Aglycone: Sapogenin units

The non-sugar unit of the saponin molecule is called aglycone, sapogenin, or genin. Depending on the sapogenin present in the molecule, the saponins can be categorized into three major classes: triterpenoid glycosides, steroid glycosides, and alkaloid steroid glycosides (Figure 6.2).

Triterpenoid glycosides are the widely distributed saponin class in the Kingdom Plantae. The triterpenes consist of 30 carbon atoms in 6 isoprene units (C_5) and polycyclic in nature. α-Amyrin, β-amyrin, and lupeol are the common pentacyclic triterpenoid sapogenin nucleus found in nature. Medicinally and widely distributed important triterpenoid glycosides contain β-amyrin backbone. Very few tetracyclic triterpenoid sapogenins have been reported [11, 27]. The triterpenoid glycosides contain hydroxyl group (-OH) at C-3, hydroxymethyl group (-CH$_2$OH) at C-26, and carboxyl group (-COOH) at C-28 (Figure 6.3). According to the attachment of the glycone sugar units to the aglycone

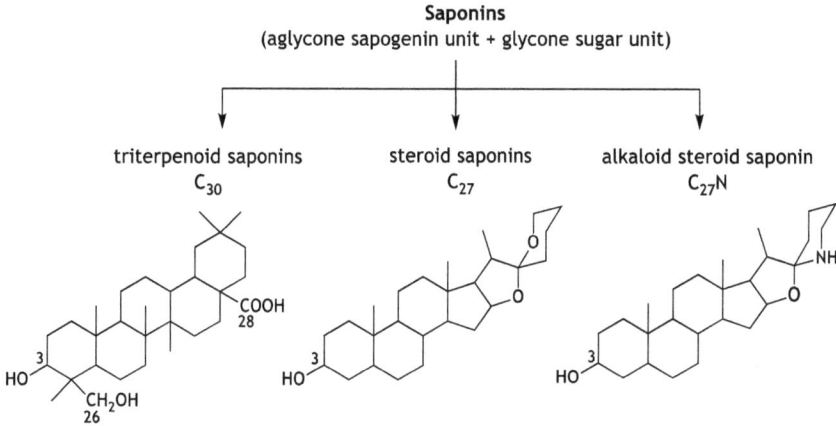

Figure 6.2: Classification of sapogenines.

nucleus, triterpenoid saponin can be categorized into mono-, bi- and tri-desmosidic saponins (in Greek, *desmos*, means chains). Monodesmosidic saponins have one sugar moiety attached to the hydroxyl group at C-3. Bidesmosidic saponins have two sugar moieties attached to the aglycone nucleus through ether linkage to C-3 position and ester linkage to C-28 (triterpene saponins) or an ether linkage to C-26 (furostanol saponins) positions. As the name implies tridesmosidic saponins have three sugar units and rarely found. Recently tridesmosidic saponins were reported in dicot (dicotyledonous) plant families commonly in Amaranthaceae, Compositae, and Fabaceae [12, 13, 27]. The feasible hydrolysis of the esterified sugar unit at C-28 resulted in monodesmosidic saponins from bidesmosidic saponins with reduced or low biological properties. Triterpenoid saponins generally occur in dicot families: Araliaceae, Chenopodiaceae, Euphorbiaceae, Hippocastanaceae, Fabaceae, Ranunculaceae, Symplocaceae, Theaceae, and Verbenaceae [7, 9, 10, 27]. In general, glycosylated triterpenes widely found in dicotyledonous plant families and very rarely in monocotyledons (monocots).

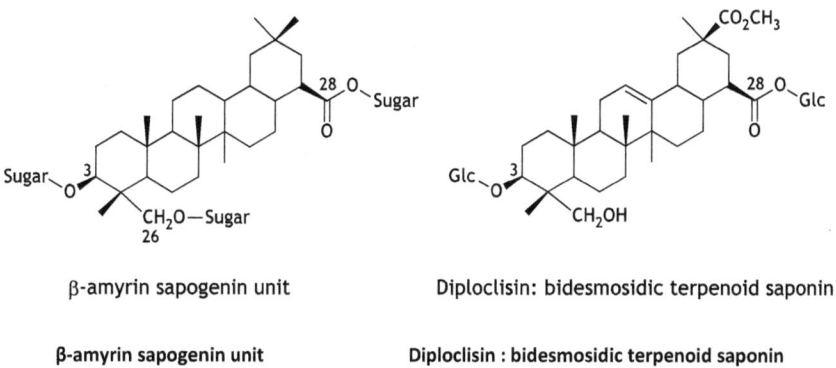

β-amyrin sapogenin unit

Diploclisin: bidesmosidic terpenoid saponin

Figure 6.3: Triterpenoid saponin glycosides.

Steroidal glycosides are modified triterpenoids of the polycyclic molecule with 27 carbon atoms with the backbone of cholesterol. The 27 carbons are distributed in the tetracyclic six-membered and bicyclic five-membered ring system. Steroidal glycosides contain two rings with a hetero molecule (a five-membered pyran ring and a six-membered furan ring). Both hetero rings have a common spiro carbon atom (C-22). The sugar unit attached to the sapogenin ring through the C-3 ether linkage makes steroidal glycoside saponins. Steroidal saponins are subdivided into two groups: spirostanol and furostanol saponins (Figure 6.4). Spirostanol saponins are characterized by the presence of a bicyclic spiroacetal moiety at position C-22 that involves in the E and F rings. Meanwhile, furostanol saponin possesses an O-linked sugar residue attached at the hydroxyl group at C-26 and spiroketal at C-22 positions. Attachment of sugar unit at C-26 in furostanol prevents the cyclization and the formation of the F ring as seen in spirostanols.

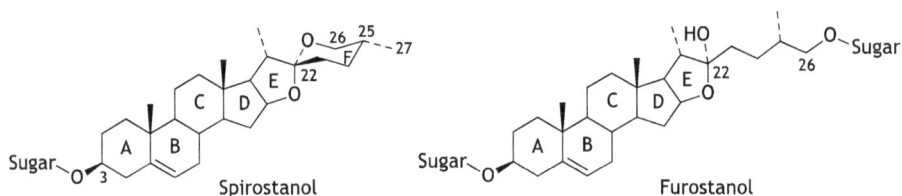

Figure 6.4: Types of steroid glycosides.

Steroidal saponins are not widely distributed in nature as the triterpenoid saponins and exhibit various biological activities. They are found abundantly in monocot families/subfamilies such as Agavaceae, Alliaceae, Dioscoreaceae, Liliaceae, Poaceae as well as in dicot families like Fabaceae, and Solanaceae. Both triterpenoid and steroidal saponins are derived from the same precursor oxidosqualene with 30 carbons. But the difference between these two classes of compounds is that the steroidal saponins containing only 27 carbons by removing three methyl groups [7, 10].

As the name implies, **alkaloid steroid glycosides** contain nitrogen in the steroidal sapogenin moiety (Figure 6.5). These compounds are toxic and commonly found in the Solanaceae family. This N containing six-membered ring is a piperidine ring. Similar to the steroidal saponins, alkaloid steroid saponins have no carboxylic acid group (-COOH) and the attachment of the sugar moiety is through the hydroxyl group at C-3 ether linkage. All the alkaloid sterol glycoside compounds have the same stereochemistry at the C-25 methyl group, i.e., equatorial methyl group and isomers at C-22. The N atom can be either secondary or tertiary. Recently, monodesmosidic and bidesmosidic types of steroid saponins and alkaloid steroid saponins have been rarely reported [14–16].

Table 6.1 compares the structural differences of triterpenoid, sterol, and alkaloid sterol saponin glycosides.

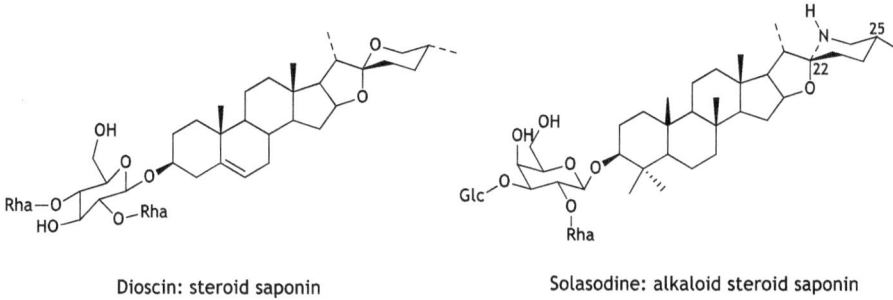

Dioscin: steroid saponin   Solasodine: alkaloid steroid saponin

Figure 6.5: Steroid and alkaloid steroid saponin glycosides.

6.2 Role of saponins on plants and microorganisms

Saponins are glycosides with a characteristic foaming ability. They are bitter and therefore, reduce palatability. The saponins act as deterrents, toxins, and digestibility inhibitors to protect plants against herbivores. The mechanism is not known yet, but it is
associated with their hemolytic activity. Here the saponin compounds enlarge the
membranes of the respiratory and digestive tract, and the sapogenins in certain saponins increase the permeability of the membranes of red blood cells. In severe cases, the
membranes are destroyed and their hemoglobin escapes into the bloodstream. This hemolytic effect varies considerably between different plant species [17–20].

For insects, it is hypothesized that steroidal saponins have close interaction with
insect molting hormones [20]. Some saponins reduce the intake and growth rate of
non-ruminant animals. For example, the saponins found in oats (Poaceae) and spinach (Amaranthaceae) increase and accelerate the absorption ability of ions (Ca^{2+}),
thus supporting the digestion process. Some weeds contain significant quantities of
toxic saponins. As an example, the legume alfalfa (*Medicago sativa*) is a high-protein
legume that is used in swine and poultry industries. But their consumption is reduced
due to the presence of toxic saponin: medicagenic acid which is responsible for its
antinutritional effects. Furthermore, the saponins present in corn cockle (*Agrostemma
githago*), soapwort (*Saponaria officinalis*), cow cockle (*Vaccaria hispanica*), and broom-
weed (*Gutierrezia sarothrae*) cause serious problems for grazing animals by weight
loss, diarrhea and gastroenteritis, abortion and listlessness [18, 20].

The distribution and localization of saponins in root systems found abundantly
as a defense mechanism to protect young plants from microbial attacks. These
types of chemical compounds known as phytoprotectants. For example, the major
saponin from *Avena* sp. (oats), avenacin A-1, is localized in the epidermal cell layer
of oat root tips and also in the lateral root initials. These saponins act as chemical
barriers to invading soil-borne microbes that attack plant tissues [21]. These saponins in the outermost epidermal cell layer act as the first barrier for the pathogenic

Table 6.1: Structural differences of saponin glycosides.

Triterpenoid glycosides	Sterol glycosides	Alkaloid sterol glycosides
Sapogenin contain C-30 β-amyrin backbone	C-27 cholesterol backbone	C-27 cholesterol backbone
Oxygen is the only heteroatom present in the sapogenin unit	Oxygen (2) is present	Oxygen and nitrogen are present
Heteroatom is not involved in the sapogenin ring system	Heteroatoms (O) are involved in five-membered pyran ring and six-membered furan ring	N is attached to six-membered furan ring (piperidine) and O is attached to five-membered pyran ring
Sapogenin contain 5-six membered rings (total of 5 rings)	4-six membered and 2-five membered rings (total of 6 rings) *spirostanol*	4-six membered and 2-five membered rings (total of 6 rings) *spirostanol*
	3-six membered and 2-five membered rings (total of 5) *furostanol*	3-six membered and 2-five membered rings (total of 5) *furostanol*
No spiro centers in the sapogenin	C-22 is a spiro center	C-22 is a spiro center
No methyl group at C-25	Methyl group at C-25 is a stereoisomer (spirostanol)	Methyl group at C-25 is always in the equatorial position
Glycone sugar is attached via OH group at C-3; COOH group at C-28 or OH group at C-26	Glycone sugar is attached via OH group at C-3; No COOH group	Glycone sugar is attached via OH group at C-3; No COOH group

microbes to invade the plant tissues. Here the steroidal saponins are stored in plant vacuole as an inactive bidesmosidic form. When the pathogenic fungi damage the plant tissues bidesmosidic saponins are activated and hydrolyze the D-glucose unit by β-glucosidase enzymes forming toxic monodesmosidic saponins [22]. This active form of saponins disrupts the fungal plasma membrane by forming membrane pores in association with fungal sterol: ergosterol, eventually, which causes fungal cell death [23]. Similarly, the saponins stored in phloem attacks the insects or root-knot nematodes that feed on phloem sap [24]. Though the saponins are sensitive to the pathogenic microbes and attacking animals, they are sensitive for the plants too. They show phytotoxicity for the growth of radicals and hypocotyls. It is known that a high dose of medicagenic acid saponins inhibits plant growth. The structure reactivity relationship studies on saponins against phytotoxic activity show that monodesmosidic saponins are more active than the bidesmosidic saponins [25, 26].

6.3 Biosynthesis of saponins

Both sterol and terpenoid saponins are the largest and structurally diverse natural products derived from mevalonate pathway. The special feature of these compounds that biosynthesized in this pathway is derived from C_5 isoprene units (fundamental building block) connected in head-to-tail combinations. Cyclization, rearrangements, loss of carbons, different combinations of isoprene units (tail-to-tail and/or head-to-head), transamination, and glycosylation reactions result in a structurally diverse array of natural saponins. The fundamental isoprene units are derived from the condensation reactions of three molecules of acetyl coenzyme A to form mevalonic acid. Mevalonic acid is the precursor of isoprene units. 3-Isopentenyl pyrophosphate (IPP) (C_5) and its isomer dimethylallyl pyrophosphate (DMAPP) (C_5) are the two common forms of isoprene units. IPP is isomerized to DMAPP by isomerase enzyme stereospecifically, this isomerization is in equilibrium and mainly favors the formation of DMAPP (Figure 6.6). Triterpenes and sterols are made of six isoprene units (C_5) and squalene (C_{30}) is the first common precursor.

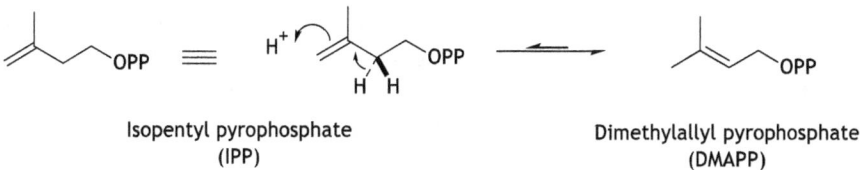

Isopentyl pyrophosphate
(IPP)

Dimethylallyl pyrophosphate
(DMAPP)

Figure 6.6: Isoprene units.

In the cytoplasmic matrix of the cell (cytosol), one molecule of IPP condensed with its isomer DMAPP to form a monoterpene called geranyl pyrophosphate (GPP, C_{10}) and further condensation with one molecule of IPP to form a sesquiterpene (C_{15}) called farnesyl pyrophosphate (FPP, C_{15}) by the enzyme prenyltransferase. The stereospecific enzymatic reactions catalyze the head-to-tail condensations of the isoprene units. Then two molecules of FPP condense to form triterpene squalene by the enzyme squalene synthase. For this condensation one molecule of NADPH is involved and eliminated as NADP by supplying hydride. The conversation of squalene-2,3-epoxide: the cyclization form of squalene, by the squalene epoxygenase enzyme, involves one molecule of NADPH and O_2. Up to this point i.e., formation of squalene-2,3-epoxide, the biosynthesis represents the same steps for triterpenoid, steroid, and steroid alkaloid sapogenins.

The protonation of epoxide group will generate a tertiary carbocation, and the electrophilic addition reaction to a double bond form six-membered ring. This process continues by generating carbocations after each ring formation until tertiary protosteryl cation is formed (Figure 6.7). The stereochemistry of the tertiary cation is controlled by the enzymetic action. This cation will undergo a series of Wagner-Meerwein rearrangements (1,2-hydride and 1,2-methyl shifts) to synthesize lanosterol and cycloartenol. The biosynthesis of cholesterol in algae and green plants use cycloartenol, while for fungi and non-photosynthetic organisms use lanesterol as the intermediate. Cycloartane triterpenes are the most abundant terpene saponins in higher plants.

Furthermore, squalene-2,3-epoxide is folded into another form of confirmation by the action of cyclase enzyme to form the dammarenyl cation. Though the both protosteryl and dammarenyl cations are synthesized by the squelene epoxide their stereochemical properties are different. This dammarenyl cation undergoes further carbocation promoted cyclizations especially by relieving the ring strain of the D ring from five-membered to six-membered forming baccharenyl cation. Furthermore, carbocation directed ring expansions give lupenyl, olenyl, and taraxasteryl cations to yield lupeol, β-amyrin, and α-amyrin respectively. These pentacyclic triterpenoid skeletons are commonly found in triterpenoid sapogenins (Figure 6.7).

The steroidal sapogenins are C_{27} sterols in which the side chain of cholesterol had modified to produce a spiroketal or spiroacetyl center. The biosynthesis of furostanol and spirostanol sapogenins is believed to be from a hypothetical cholesterol intermediate (16,26-dihydroxy-22-oxo-cholestan) Alkaloid sterol saponins are nitrogen analogues of steroid saponins. Here the nitrogen atom is introduced by the transamination reaction using arginine as the precursor amino acid. However, natural steroid sapogenin and alkaloid steroid sapogenins are very closely related and share the same biosynthesis and metabolism (Figure 6.8) [3, 7, 10, 27–30].

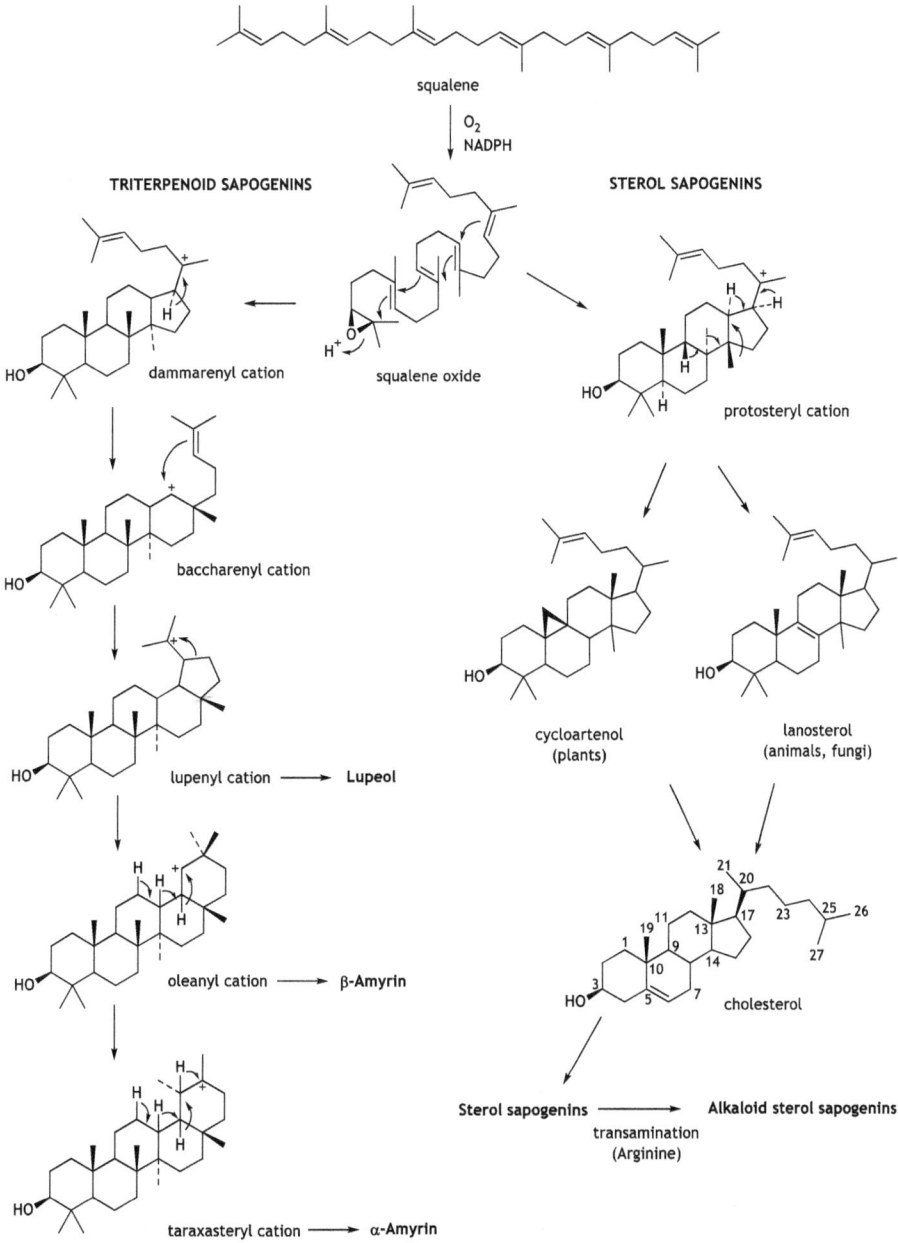

Figure 6.7: Biosynthesis of triterpenoid, steroid, and alkaloid sterol sapogenins.

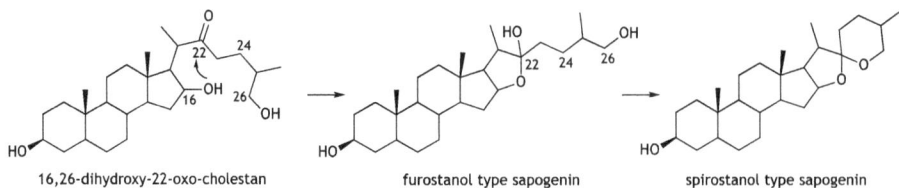

Figure 6.8: Biosynthesis of furostanol and spirostanol sapogenins.

6.4 Pharmacologically important saponins from plants

Epidemiological research studies suggested that diet is one of the major important environmental factors contributing to the etiology of the most predominant forms of non-communicable diseases such as diabetes, cancer, hyperglycemia, cardiovascular disease, and hyperlipidemia. Plant food sources contain macronutrients as well as a wide range of micro components such as enzyme inhibitors and secondary metabolites, saponins, alkaloids, and flavanones. These microcomponents are named "nutraceuticals" or "phytochemicals," which are considered as non-essential micronutrients possessing a vital role in maintaining human health [31, 32]. These microcomponents are recognized as biologically active secondary metabolites of the plant food sources. Their contribution to the prevention of chronic diseases/non-communicable diseases is currently being intensively studied [31, 33]. Biological activity of saponin is being explored in a large number of research groups in the world with the hope of identifying novel bioactive compounds as drug targets to combat non-communicable diseases. In order to restrict the usage of conventional medicine for non-communicable diseases, there is a growing trend of consuming food plants that are rich in biologically active secondary metabolites. Saponins are an important group of secondary metabolites recognized with a wide range of pharmacological and cosmetic applications.

Saponins, which are composed of a sugar moiety linked to a hydrophobic aglycone (sapogenin), are widely distributed plant natural products with massive structural and functional diversity. They are the key ingredients of many herbal drugs employed in phytotherapy, traditional herbal cosmetics, and in folk medicines. Currently, commercially available crude drugs produced using seeds, leaves, stem bark, roots, and rhizomes of higher plants contain a considerable amount of saponins which are responsible for the efficacy of the crude drugs [34]. The amphiphilic nature of saponins, due to their lipophilic aglycones linked to hydrophilic carbohydrate/sugar side chains, is responsible for the surface-active and numerous pharmacological properties of saponins. The wide range of biological activities of saponins depends on the source and unique chemical structure [31]. As some form of saponins are the starting material for the semi-synthesis of steroidal drugs, they have also been used in the pharmaceutical industry. In consideration of the physiological, immunological, and pharmacological properties of saponins, clinical studies are also being carried out [35].

Food sources rich in saponins have traditionally been known as "antinutritional factors" while some food sources have a limited consumption due to bitter taste. As ongoing scientific studies proved the health benefits of plants containing saponins, consumption of food and non-food sources of saponins is being increased significantly. Recent research showed that the active chemical constituent of many herbal medicines/foods which contribute to the relevant health benefits are saponins. For example, most commonly used food sources such as soybeans (*Glycine max*) and garlic (*Allium sativum*) are rich in biologically active saponins. A single plant species contains a chemically diverse mixture of saponins and showed a wide range of pharmacology activities [34]. Moreover, the key ingredients of traditional Chinese medicine are polyphenols and saponins [36].

A considerable quantity of saponins is present in many food plants. The predominant food sources which contain saponins are legumes: chickpeas (*Cicer arietinum*), soybeans (*G. max*), broad beans (*Vicia faba*), kidney beans (*Phaseolus vulgaris*), peanuts (*Arachis hypogaea*), and lentils (*Lens culinaris*). The other dietary sources which are rich in saponins are tea (*Camellia sinensis*), oats (*Avena sativa*), asparagus (*Asparagus officinalis*), sugar beet (*Beta vulgaris*), spinach (*Spinacia oleracea*), *Allium* spp., and yams. The major non-food saponins rich sources are soap bark tree (*Quillaja saponaria*), ginseng (*Panax* spp.), fenugreek (*Trigonella foenum-graecum* L.), alfalfa (*Medicago sativa*), soapwort (*Saponaria officinalis*), licorice (*Glycyrrhiza glabra*), Mojave yucca (*Yucca schidigera*), *Gypsophila paniculata*, *Smilax regelii*, and horse chestnut (*Aesculus hippocastanum*) are used in medicinal and industrial applications [34, 37]. Saponins with complex structural and functional diversity have been reported from soybeans and ginseng [38, 39].

Ginseng (*Panax* spp.), a functional food and health-enhancing supplement, has been used in East Asia for thousands of years. Recently, many clinical trials have been carried out to evaluate the pharmacological properties of ginseng in Western countries as well as in Eastern countries. More than 100 different saponins with different pharmacological activities such as antitumor, antioxidant, antifatigue, neuroprotective, and osteoclast genesis inhibitory effects have been reported from the root of *Panax ginseng* (red ginseng) [40]. Major biological active saponins in *P. ginseng* are called ginsenosides and comprise triterpenoid dammarane structures. More than 30 ginsenosides have been identified from ginseng which are beneficial for conditions like diabetes mellitus, cancer, as well as disorders of the cardiovascular system and immune system [41].

Fenugreek (*Trigonella foenum-graecum* L.) has long been utilized as a traditional medicine possessing restorative properties as well as pungent aromatic properties; hence, it is used as a spice for curry preparation in Asian and Mediterranean countries. Furostanol-type saponins (**1**), isolated from fenugreek, increased food consumption, induced hyperinsulinemia, and reduced plasma total cholesterol levels without change in triglycerides [42]. The crude drug was prepared by using dry roots of *Bupleurum frutescens* which contained saikosaponins. This crude drug is traditionally used in the treatment of inflammation and it is listed in Chinese and Japanese pharmaco-poeias [43]. Moreover, the roots of *Chlorophytum borivilianum* are extensively utilized in many therapeutic applications in the traditional medicine system in India for diabetes, arthritis, and increasing general body immunity. It possesses sper-matogenic property and is used to cure impotency. Dried roots of *C. borivilianum* contain 2–17% of saponins that have been used as an alternative to "Viagra" [44]. Another important plant material, licorice, the root of *Glycyrrhiza glabra*, *G. uralensis*, and their varieties belong to family Fabaceae, is utilized as medicine since ancient times in Western and Eastern countries. Licorice has been utilized as an expectorant, antitussive agent and sweetening agent in Western countries while it has been de-scribed as a drug for increasing physical strengthening and curing wounds in China. It is also used as an antidote and to treat throat and skin inflammation. It contains saponin, particularly glycyrrhizin (**2**) which is sweet in taste, exhibiting low hemo-lytic index and is clinically effective for the treatment of gastric ulcers [45].

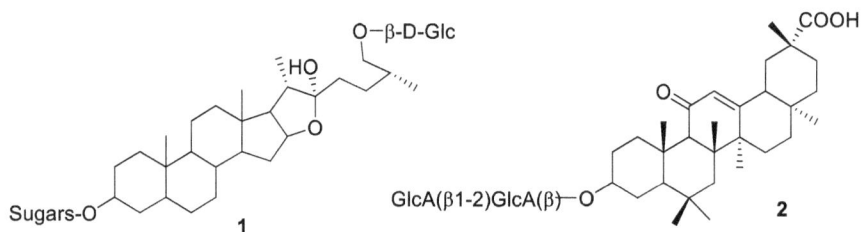

Saponins have shown a wide range of biological activities and are thus used as tra-ditional medicine for a long period; in addition, they have also been found to affect the growth and reproduction in animals [46]. Some of the important biological ac-tivities of saponins are discussed in detail below.

6.4.1 Hemolytic properties

It is revealed that plants containing saponins possess hemolytic properties toward red blood cells. For examples, the seeds of *Barringtonia asiatica*, which are known to contain saponins, have been utilized to catch fish by native Asian and Pacific fisher-men from ancient times. The deadly effect of saponins on cold-blooded animals is documented comprehensively in Australian history by Aboriginal cultures who used saponins-containing plant materials as fish poisons [38, 44]. Saponins possess hemo-lytic properties which are generally attributed to the impairment of erythrocyte mem-brane leading to rupture of erythrocytes. Hemolytic action is triggered by the affinity of the aglycone moiety for membrane sterols [35]. It is reported that in contrary to free aglycone, all oleanolic acid glycosides including Glucuronide F (**3**) and Glucuronide D₂ (**4**) isolated from *Calendula officinalis* are hemolytic agents [47]. The level of hemolytic activity depends on the type of aglycone and the sugar side chains of the saponin molecule. Saponin mixture isolated from *Maesa lanceolata* Oleanolic saponin mixture showed high hemolytic activity [44].

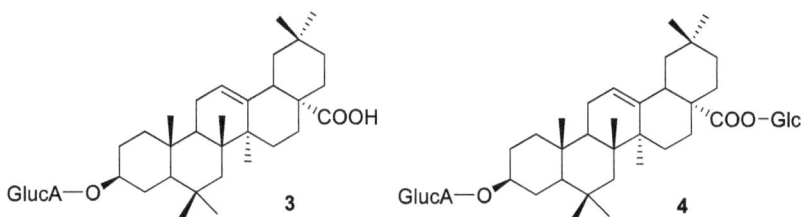

6.4.2 Anticarcinogenic activities

Saponins possess anticarcinogenic activity via numerous mechanisms. The cytotoxic effect of saponins against tumor development has been evaluated by *in vitro* and *in vivo* studies. The key ingredients in several herbal medicines which are used as chemotherapeutic agents are saponins. Chinese herbal drug Yunan Baiyao showed cytotoxic activity in several cancer cell lines [31]. The root of *P. ginseng* has been widely used in some Asian and Western countries as a promising remedy with cancer-preventive effects and was found to possess cytotoxic and anti-metastatic activities against numerous types of cancer cell lines [48]. Ginsenosides Rh2 (**5**) and Rg3 (**6**), which are constituents of red ginseng, inhibited the proliferation of prostate cancer cells, while Ginsenoside Rh 2 (**5**) inhibited *in vitro* human ovarian cancer cell (HRA) proliferation in a dose-dependent manner at concentration ranged from 10 to 100 μL. In addition, ginsenoside Rh2 (**5**) showed a reduction of cell proliferation and increasing of sub-G1 cells in two cultured intestinal cell lines: Int-407 and Caco-2 [49].

Steroidal glycosides have been identified from different species of *Allium* and the tumor-inhibitory effects of these compounds have been studied using several experimental models. For example, several spirostanol glycosides isolated from different *Allium* species displayed fairly high cytotoxic activity on promyelotic leukemia cells HL-60. The spirostanol saponin eruboside-B (**7**) isolated from *A. leucanthum* exhibited *in vitro* cytotoxic activity against A549 WS1, and DLD-1 cells [50]. Similarly, tubeimoside-1 was isolated as a major bioactive constituent of the traditional Chinese medicinal plant *Bolbostemma paniculatum*, which is widely utilized for the cure of tumors. This compound exerts anticancer activity by the induction of apoptosis, arresting cell cycle, and inhibiting metastasis by specifically targeting multiple signaling pathways which are generally deregulated in various cancers [51]. Moreover, saponin fractions from *Crocosmia crocosmiiflora* displayed significant anti-tumor activities against Ehrlich ascites carcinoma in Jd-ICR mice, glycyrrhizin (**2**) exhibited antimutagenic activity, while soyasaponin I from *Wisteria brachybotrys* remarkably inhibited mouse skin tumor promotion [52].

Another example of anticarcinogenic properties of saponins is the crude extract of *Hedera helix* which has exerted a cytotoxic activity on Ehrlich tumor cells, both *in vitro* and *in vivo*. Chemical constituents isolated from this plant, α-hederin (**8**) and β-hederin (**9**), exhibited cytotoxicity as well as antimutagenic properties [45]. A preliminary screening confirmed that α-hederin (**8**) has cytotoxic effects on mouse B 16 melanoma cells and in 3T3 mouse fibroblasts at relatively low concentrations [52].

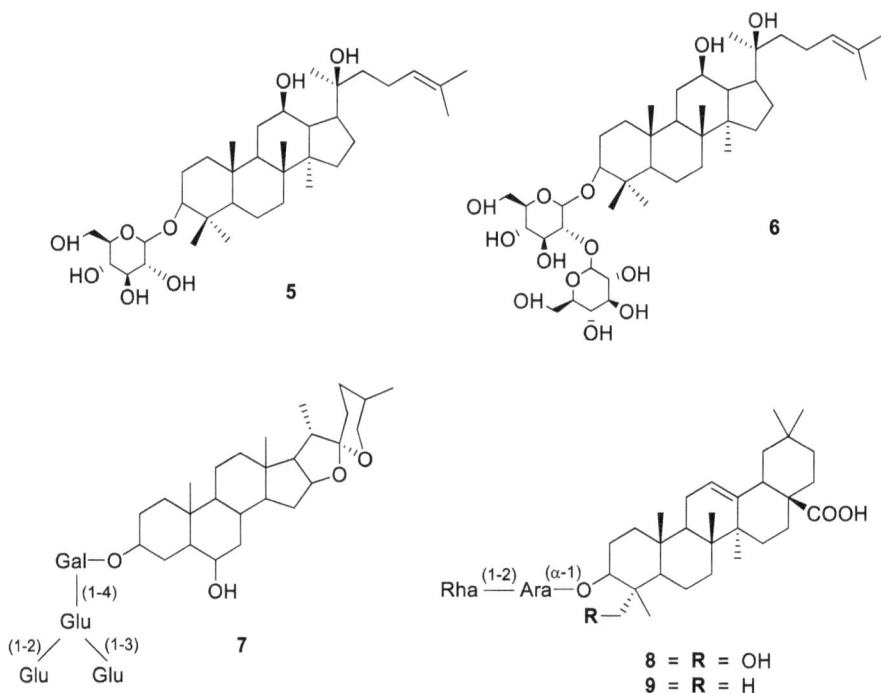

6.4.3 Molluscicidal activity

Schistosomiasis is also known as snail fever caused by a parasite, *Schistosoma*, which is served by intermediate hosts of certain species of aquatic snails. *Bulinus* and *Biomphalaria*, which cause urinary bilharzia, act as intermediate hosts in the life cycle of *Schistosoma*. A high prevalence rate of schistosomiasis has been reported in African, Asian, and South American countries. Due to the increasing costs of synthetic molluscicides, plants that could kill snails have drawn attention as less expensive and environmentally friendly alternatives. Saponins that are present in those plants are highly toxic to molluscs and have been investigated as molluscicides for the control of schistosomiasis [53]. Most saponins are toxic to cold-blooded species, whereas they exert only weak toxicity on warm-blooded species upon oral administration, which might be attributed to low absorption rates. Molluscicidal activity against *Pomacea canaliculata* was observed in monodesmosidic saponin, 3-*O*-β-D-glucopyranosyl- (1→3)-α-L-arabinopyranosyl phytolaccagenic acid (**10**) [54, 55]. Two of the saponins isolated from *Swartzia simplex* exhibited a very high molluscicidal activity against the schistosomiasis-transmitting snail *Biomphalaria glabrata*. Three saponins, 3-*O*-β-D-glucopyranosides of hederagenin (**11**), bayogenin (**12**) and medicagenic acid (**13**), isolated from roots of *Dolichos kilimandscharicus* exhibited molluscicidal activity against *B. glabrata* [56]. Moreover, a six-oleanane-type triterpenoid saponin mixture isolated from *Maesa lanceolata* was tested for molluscicidal activity against *B. glabrata* with LD_{95} and LD_{50} values of 4.1 and 2.3 µg/ml, respectively [57].

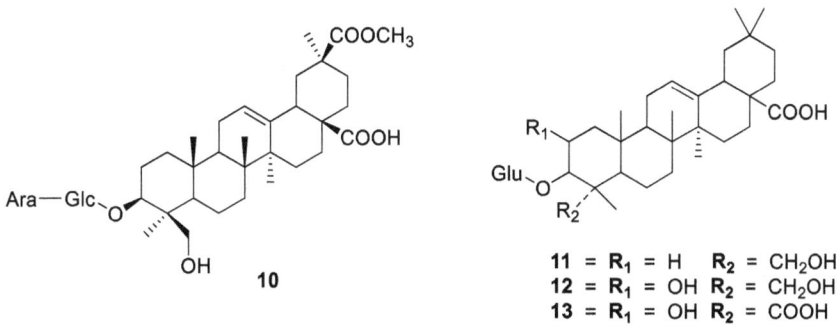

11 = R_1 = H R_2 = CH_2OH
12 = R_1 = OH R_2 = CH_2OH
13 = R_1 = OH R_2 = COOH

6.4.4 Vaccine adjuvant and immunostimulant

An adjuvant can be used for improving the response of immunogenicity of weak antigen, enhancing the effectiveness of vaccine and decreasing the amount of antigen/immunizations [58]. New generations of vaccines are based on the recombinant proteins and DNA; therefore, their reactogenicity and immunogenic properties are probable to be less than vaccines currently used. Hence, there is a requirement for the improvement of new vaccine adjuvant. Though a wide range of adjuvants have been utilized as experimental vaccines trials, most of these materials have limited their potential applications due to undesirable side effects. One of the well-known steps undertaken to address this issue was the identification of saponins from *Quillaja saponaria* with adjuvant activity. The most active fractions isolated from *Q. saponaria* as vaccine adjuvants were QuilA and QS-21 (**14**). Quil A, which is composed of more than 23 different saponins, has been used successfully for veterinary applications. QS-21 (**14**) has been assessed as an adjuvant for DNA vaccines and also showed significant dose reduction during HIV-1 envelope subunit immunization in humans. Nevertheless, there are serious drawbacks associated with the use of these active compounds QuilA and QS-21 as a vaccine adjuvant, particularly, the high toxicity and undesirable hemolytic effects, thus limiting their usage in human vaccination [59]. These saponins alone or incorporated into immunostimulating complexes (ISCOMs) are utilized as adjuvants in commercially available some veterinary vaccines and have been studied as adjuvants in human experimental vaccines [60–62].

Apart from *Q. saponaria*, saponins isolated from several other plants have also displayed immunostimulator properties. The sponins from the root of *Achyranthes bidentata* was reported as immunostimulator while the Chinese traditional herb *Astragalus* which is believed to strengthen and boost the immune system was reported with triterpene saponins like astragalosides I–X, isoastragalosides I–IV and soyasaponin I. Moreover, the chromatographic separation of root extract of *Panax notoginseng* afforded seven adjuvant active protopanaxatriol-type saponins while the oral administration of ginseng extract was reported with enhanced antibody response and blood lymphocyte proliferation in human [59, 60].

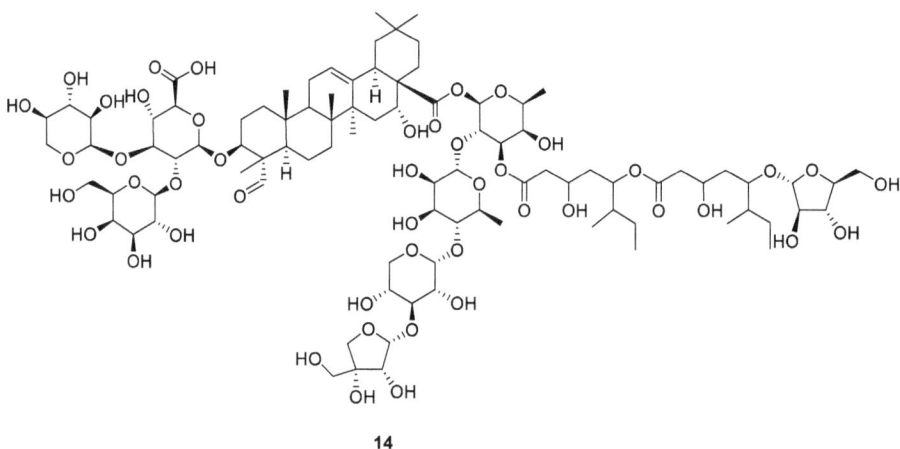

14

6.4.5 Antiviral activity

Rosamultin (**15**) isolated from *Sargentodoxa cuneate* and TS s21 (**16**) isolated from *Thinouia coriacea* showed antiherpetic (HSV-1) activity with EC_{50} values of 25 and 2.7 µM respectively. Rosamultin (**15**) inhibited the viral capsid protein synthesis of herpes simplex virus type 1, while TS s21 (**16**) inhibited herpes simplex virus type 1 DNA synthesis [63].

15   **16**

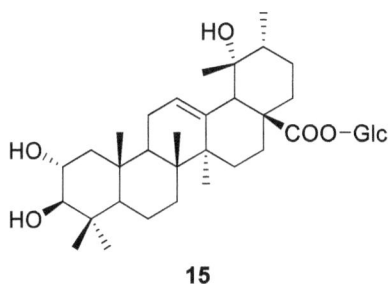

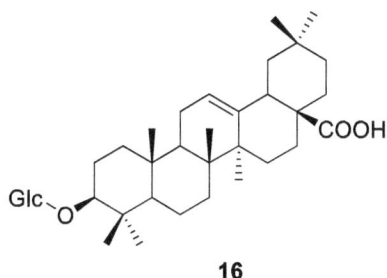

6.4.6 Hypocholesterolaemic activity

Experiments with animal models revealed that plants containing saponins possess the ability to inhibit cholesterol absorption from the intestinal lumen, and, consequently, reduction of the concentration of plasma cholesterol. This may be due to saponins combine with cholesterol in the digestive tract and formed complex formation or plant saponins directly involve on cholesterol metabolism. Particularly, the cholesterol-lowering effect of garlic (*Allium sativum*) is attributed to the presence of steroid saponins. A low plasma total and LDL (low-density lipoprotein) cholesterol concentrations without any change in the HDL (high-density lipoprotein) cholesterol level was observed in rats fed with saponin-rich fraction from raw garlic at 10 mg/kg/day [50].

6.4.7 Regulation of blood glucose concentration

In vitro and *in vivo* studies revealed that the root of *P. ginseng* and other ginseng species possess anti-hyperglycemic activity and increases glucose homeostasis [40]. Pharmacological studies demonstrated that Ginsenoside Rb2 (**17**) was highly effective with its ability to significantly decrease the blood glucose level with increased regulation of glucokinase and glucose 6-phosphatase.

In addition, anti-hyperglycemic and antiobesity effects of *P. ginseng* berry extract were identified and its major constituent, Ginsenoside Re (**18**), was found to be the active compound. The active compound isolated from the American ginseng (*P. quinquefolius*), Ginsenoside Rb1 (**19**), was found to enhance glucose-stimulated insulin secretion [40, 49].

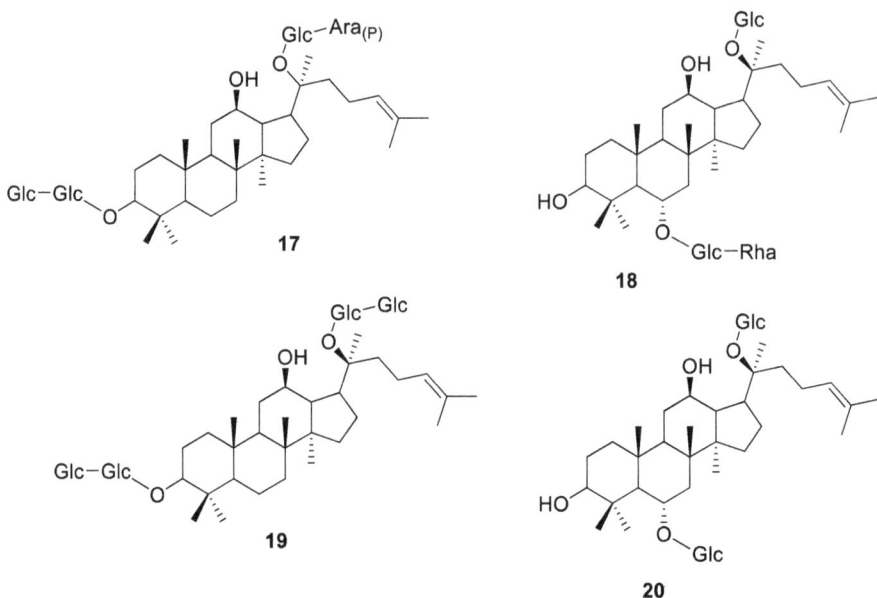

Ginsenoside Rg1(**20**) exhibited normalizing of blood pressure and also ginseng can be used as in treatment of pulmonary and systemic hypertension. Furthermore, Ginsenosides Rg1 and Rg3 isolated from root of *P. ginseng* exhibited anti-platelet and anti-atherosclerotic properties that may be used to prevent and treat for certain thrombotic and atherosclerotic disorders [49].

6.4.8 Antifungal activity

Antifungal activity of saponins has been known from ancient times. It is revealed that tigogenin saponins, **21**, **22**, **23**, and **24** with a sugar moiety of four or five monosaccharide units possess remarkable activity against *Candida neoformans* with minimum fungicidal concentration (MFC) comparable to the positive control amphotericin B as well as significant activity against *Aspergillus fumigatus* without exhibiting cytotoxic activity on mammalian cells [64]. Moreover, glucuronide F (**3**) and glucuronide D_2 (**4**) isolated from *Calendula officinalis* showed antifungal activity against *Trichoderma viride* [47] while eruboside-B (**7**) from garlic (*Allium sativum*) exhibited antifungal activity against *C. albicans* with a MIC of 25 mg/mL [65]. Quinoa saponins isolated from *Chenopodium quinoa* were found to have significant antifungal activity; the crude saponin mixture showed activity against *C. albicans* with a MIC of 50 µg/mL; however, the purified individual compounds of saponins exhibited significantly little or no activity, suggesting some possible synergistic interactions between saponin mixture [55].

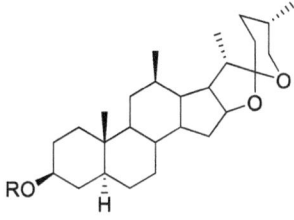

21 = **R** = Gal(4-1)-Glc[(3-1)Xyl](2-1)Glc(3-1)Rha
22 = **R** = Gal(4-1)-Glc[(3-1)Glc](2-1)Glc(3-1)Rha
23 = **R** = Gal(4-1)-Glc[(3-1)Xyl](2-1)Glc
24 = **R** = Gal(4-1)-Glc[(3-1)Xyl](2-1)Glc(3-1)Xyl

6.5 Conclusion

Saponins are widely distributed among plants and are important in human and animal nutrition. The biological properties of plants that are rich in saponins have been utilized since ancient times and currently, extensive research is ongoing to scientifically validate these bioactivities.

References

[1]   Riguera R Isolating bioactive compounds from marine organisms. J Mar Biotechnol 1997, 5(4),187–193.
[2]   Yoshiki Y, Kudou S, Okubo K Relationship between chemical structures and biological activities of triterpenoid saponins from soybean. Biosci Biotechnol Biochem 1998, 62(12), 2291-2299.
[3]   Desai SD, Desai D, Kaur H Saponins and their biological activities. Pharma Times 2009, 41, 13–16.
[4]   Rohit S, Nidhi S, Gulab TS, Bhagwan SS, Pallavi J Conventional method for saponin extraction from *Chlorophytum borivilianum* Sant. et Fernand. Global J Res Med Plants & Indigen Med 2014, 3(2), 33–39.
[5]   Moghimipour E, Handali S Saponin: Properties, methods of evaluation and applications. Annu Res Rev Biol 2014, 5(3),207–220.
[6]   Khan MMAA, Naqvi TS, Naqvi MS Identification of phytosaponins as novel biodynamic agents: An updated overview. Asian J Exp Biol Sci 2012, 3(3),459–467.
[7]   Hostettmann K, Marston A Saponins. Cambridge, UK, Cambridge University Press, 1995.
[8]   Kochetkov NK, Khorlin A, Chirva VJ Clematoside C-triterpenic oligoside from *Clematis manshurica*. Tetrahedron Lett 1965, 6(26),2201–2205.
[9]   Güçlü-Üstündağ Ö, Mazza G Saponins: Properties, applications and processing. Crit Rev Food Sci Nutr 2007, 47(3),231–258.
[10] Mma EA, Ashour AS, Melad ASG A review on saponins from medicinal plants: Chemistry, isolation, and determination. J Nanomed Res. 2019, 8(1), 6–12.

[11] Basu N, Rastogi RP Triterpenoid saponins and sapogenins. Phytochemistry 1967, 6(9),1249–1270.

[12] Semmar N, Tomofumi M, Mrabet Y, Lacaille-Dubois M-A (2010) Two new acylated tridesmosidic saponins from *Astralagus armatus*. Helv Chim Acta 2010, 93, 871–876.

[13] Kuljanabhagavad T, Wink M Biological activities and chemistry of saponins from *Chenopodium quinoa* Willd. Phytochem Rev 2009, 8(2),473–490

[14] Ali Z, Khan IA Monodesmosidic spirostane type steroid glycosides from *Dioscorea caucasica*. Planta Med 2012, 78(05), Pl430. doi: 10.1055/s-0032-1321117

[15] Ali Z, Smillie TJ, Khan IA Cholestane steroid glycosides from the rhizomes of *Dioscorea villosa* (wild yam). Carbohydr Res 2013, 370,86-91.

[16] Challinor VL, De Voss JJ Open-chain steroidal glycosides, a diverse class of plant saponins. Nat Prod Rep 2013,30, 429–454

[17] Wittstock U, Gershenzon J Constitutive plant toxins and their role in defence against herbivores and pathogens. Curr Opin Plant Biol 2002,5, 300–307.

[18] Osbourn A Saponins and plant defence – a soap story. Trends Plant Sci 1996, 1,4–9

[19] Dinan L, Bourne PC, Meng Y, Sarker SD, Tolentino RB, Whiting P Assessment of natural products in the *Drosophila melanogaster* B(II) cell bioassay for ecdysteroid agonist and antagonist activities. Cell Mol Life Sci 2001,58(2), 321-342

[20] Faizal A, Geelen D 2013. Saponins and their role in biological processes in plants. Phytochem Rev 2013, 12, 877–893

[21] Morrissey JP, Osbourn AE Fungal resistance to plant antibiotics as a mechanism of pathogenesis. Microbiol Mol Biol Rev 1999,63(3),708–724.

[22] Morant AV, Jorgensen K, Jorgensen C, Paquette SM, Sanchez-Perez R, Moller BL, Bak S Beta-glucosidases as detonators of plant chemical defense. Phytochemistry 2008, 69,1795–1813.

[23] Weete JD Structure and function of sterols in fungi. Adv Lipid Res 1989, 23,115–167

[24] Henry M, Rochd M, Bennini B Biosynthesis and accumulation of saponins in *Gypsophila paniculata*. Phytochemistry 1991, 30(6),1819–1821

[25] Waller GR Biochemical frontiers of allelopathy. Biol Plant 1989, 31, 418. Doi: 10.1007/BF02876217

[26] Tava A, Avato P (2006) Chemical and biological activity of triterpene saponins from *Medicago* species. Nat Prod Commun 2006, 1, 1159–1180.

[27] Sahu NP, Banerjee S, Mondal NB, Mandal D. Steroidal Saponins. In: Columbus ADK, Linz HF, Sapporo JK, ed. Fortschritte Der Chemie Organischer Naturstoffe / Progress in the Chemistry of Organic Natural Products-vol 89, Vienna, Austria, Springer-Verlag, 2008, 45–141.

[28] Thakur M, Melzig MF, Fuchs H, Weng A 2011. Chemistry and pharmacology of saponins: Special focus on cytotoxic properties. Botanics Targets Ther 2011, 1, 19–29.

[29] Netala VR, Ghosh SB, Bobbu P, Anitha D, Tartte V Triterpenoid saponins: A review on biosynthesis, applications and mechanism of their action. Int J Pharm Pharm Sci 2015, 7(1),24–28.

[30] Negi JS, Bisht VK, Singh P, Rawat MSM, Joshi GP 2013. Naturally occurring xanthones: Chemistry and biology. J Appl Chem 2013, Volume 2013, Article ID 621459, doi.org/10.1155/2013/621459

[31] Rao AV, Sung MK Saponins as anticarcinogens. J Nutr. 1995, 125(suppl_3):717S–724S.

[32] Neuhouser ML, Miller DL, Kristal AR, Barnett MJ, Cheskin LJ Diet and exercise habits of patients with diabetes, dyslipidemia, cardiovascular disease or hypertension. J Am Coll Nutr 2002, 21(5), 394-401

[33] Namdeo AG Plant cell elicitation for production of secondary metabolites: A review. Pharmacogn Rev. 2007, 1(1):69–79.

[34] Takechi M, Doi K, Wakayama Y Biological activities of synthetic saponins and cardiac glycosides. Phytother Res 2003, 17, 83–85.

[35] Francis G, Kerem Z, Makkar HP, Becker K The biological action of saponins in animal systems: A review. Br J Nutr 2002, 88(6),587–605.

[36] Liu J, Henkel T Traditional Chinese medicine (TCM): Are polyphenols and saponins the key ingredients triggering biological activities?. Curr Med Chem 2002, 9(15), 1483-1485.

[37] Milgate J, Roberts DC The nutritional & biological significance of saponins. Nutr Res 1995, 15(8),1223–1249.

[38] Yoshiki Y, Kudou S, Okubo K Relationship between chemical structures and biological activities of triterpenoid saponins from soybean. Biosci Biotechnol Biochem 1998, 62(12), 2291-2299.

[39] Park S, Ahn IS, Kwon DY, Ko BS, Jun WK Ginsenosides Rb1 and Rg1 suppress triglyceride accumulation in 3T3-L1 adipocytes and enhance beta-cell insulin secretion and viability in Min6 cells via PKA-dependent pathways. Biosci Biotechnol Biochem 2008,72(11), 2815-2823.

[40] Wang C, Liu J, Deng J, Wang J, Weng W, Chu H, Meng Q Advances in the chemistry, pharmacological diversity, and metabolism of 20(R)-ginseng saponins. J Ginseng Res 2020, 44(1),14-23.

[41] Takagi K, Saito H, Tsuchiya M Pharmacological studies of *Panax ginseng* root: Pharmacological properties of a crude saponin fraction. Jpn J Pharmacol 1972, 22(3), 339-346.

[42] Petit PR, Sauvaire YD, Hillaire-Buys DM, Leconte OM, Baissac YG, Ponsin GR, Ribes GR Steroid saponins from fenugreek seeds: Extraction, purification, and pharmacological investigation on feeding behavior and plasma cholesterol. Steroids. 1995, 60(10), 674-680.

[43] Sparg SG, Light ME, van Staden J Biological activities and distribution of plant saponins. J Ethnopharmacol 2004, 94(2–3), 219-243.

[44] Kaushik N Saponins of *Chlorophytum* species. Phytochem Rev 2005, 4, 191–196.

[45] Shibata S Saponins with biological and pharmacological activity. In: Wagner HK, Wolff PM eds, New natural products and plant drugs with pharmacological, biological or therapeutical activity. 1st ed. Berlin-Heidelberg, Germany, Springer -Verlag, 1977. 177–196.

[46] Acuña UM, Atha DE, Ma J, Nee MH, Kennelly EJ Antioxidant capacities of ten edible North American plants. Phytother Res 2002, 16(1), 63-65.

[47] Szakiel A, Ruszkowski D, Janiszowska W Saponins in *Calendula officinalis* L.–Structure, biosynthesis, transport and biological activity. Phytochem Rev 2005, 4,151–158.

[48] Park JD, Rhee DK, Lee YH Biological activities and chemistry of saponins from *Panax ginseng* CA Meyer. Phytochem Rev 2005, 4, 159–175.

[49] Popovich DG, Kitts DD Ginsenosides 20 (S)-protopanaxadiol and Rh2 reduce cell proliferation and increase sub-G1 cells in two cultured intestinal cell lines, Int-407 and Caco-2. Can J Physiol Pharmacol 2004, 82(3),183–190.

[50] Sobolewska D, Michalska K, Podolak I, Grabowska K Steroidal saponins from the genus *Allium*. Phytochem Rev 2016, 15, 1–35.

[51] Zafar M, Sarfraz I, Rasul A, Jabeen F, Samiullah K, Hussain G, Riaz A, Ali M Tubeimoside-1, triterpenoid saponin, as a potential natural cancer killer. Nat Prod Commun 2018,13(5), 643–650.

[52] Lacaille-Dubois MA, Wagner H A review of the biological and pharmacological activities of saponins. Phytomedicine. 1996, 2(4),363–386.

[53] Borel C, Hostettmann K Molluscicidal saponins from *Swartzia madagascariensis* DESVAUX. Helv Chim Acta 1987, 70(3),570–576.

[54] Borel C, Gupta MP, Hostettmann K Molluscicidal saponins from *Swartzia simplex*. Phytochemistry.1987, 26(10),2685–2689.

[55] Kuljanabhagavad T, Wink M Biological activities and chemistry of saponins from *Chenopodium quinoa* Willd. Phytochem Rev 2009, 8, 473–490.

[56] Marston A, Gafner F, Dossaji SF, Hostettmann K Fungicidal and molluscicidal saponins from *Dolichos kilimandscharicus*. Phytochemistry. 1988, 27(5),1325–1326.

[57] Sindambiwe JB, Calomme M, Geerts S, Pieters L, Vlietinck AJ, Vanden Berghe DA Evaluation of biological activities of triterpenoid saponins from *Maesa lanceolata*. J Nat Prod 1998, 61(5),585–590.

[58] Sun HX, Xie Y, Ye YP Advances in saponin-based adjuvants. Vaccine 2009, 27(12),1787-1796.

[59] Rajput ZI, Hu SH, Xiao CW, Arijo AG Adjuvant effects of saponins on animal immune responses. J Zhejiang Univ Sci B 2007, 8(3),153–161.

[60] Sun H, Yang Z, Ye Y Structure and biological activity of protopanaxatriol-type saponins from the roots of *Panax notoginseng*. Int Immunopharmacol 2006, 6(1),14-25.

[61] Sjölander A, Cox JC Uptake and adjuvant activity of orally delivered saponin and ISCOM™ vaccines. Adv Drug Deliv Rev 1998, 34(2–3):321–338.

[62] Fleck JD, Betti AH, Da Silva FP, Troian E, Olivaro C, Ferreira F, Verza SG Saponins from *Quillaja saponaria* and *Quillaja brasiliensis*: Particular chemical characteristics and biological activities. Molecules 2019, 24(1), 171. doi:10.3390/molecules24010171

[63] Simões CM, Amoros M, Girre L Mechanism of antiviral activity of triterpenoid saponins. Phytother Res 1999,13(4),323–328.

[64] Yang CR, Zhang Y, Jacob MR, Khan SI, Zhang YJ, Li XC Antifungal activity of C-27 steroidal saponins. Antimicrob Agents Chemother 2006, 50(5),1710–1714.

[65] Matsuura H Saponins in garlic as modifiers of the risk of cardiovascular disease. J Nutr 2001, 131(3),1000S–1005S.

Part III: **Specific topics**

Sathya Sambavathas, Nilupa R. Amarasinghe, Lalith Jayasinghe,
Yoshinori Fujimoto

7 Acetylcholinesterase inhibitory activity of spices and culinary herbs

7.1 Introduction

Cholinesterases play an important role in the area of neurobiology, toxicology, and pharmacology, out of which acetylcholinesterase (AChE) and butyrylcholinesterase (BChE) are vital in nerve impulse transmission [1]. The enzyme AChE catalyzes the hydrolysis of the ester bond of acetylcholine to terminate the nerve impulse transmission at the cholinergic synapses [2]. Compounds that are capable of suppressing the activity of AChE are known as AChE inhibitors or anticholinesterases. Anticholinesterases inhibit AChE enzyme, thereby increase the levels of acetylcholine near the synaptic cleft of cholinergic neurons. Prolonged availability of acetylcholine facilitates the cholinergic nerve impulse transmission process in patients with cognitive decline. AChE inhibitors are used to treat many pathological conditions, including Alzheimer's disease (AD), Parkinson's disease, myasthenia gravis, postural tachycardia syndrome, etc. AD is one of the progressive neurodegenerative diseases that affect memory and cognitive behaviour [3–4]. People of age over 60 years are affected most and it is one of the major causes of dependency among elderly people. About 5% of the world's elderly population (47 million people) was affected with dementia in 2015, and this figure is predicted to increase to 75 million in 2030 and 132 million by 2050. At present nearly 60% of people with dementia live in low- and middle-income countries [5].

There are many synthetic and natural anticholinesterases. Donepezil, tacrine, metrifonate and galantamine are anticholinesterases approved by the United States Food and Drug Administration (FDA) and currently in use. Galantamine is a plant-derived natural product [6]. In traditional medicine, many plants have been used to treat cognitive disorders. A large number of plants have been screened using well-established *in vitro* methods. Ethnopharmacological approach and bioassay-guided fractionation have facilitated the identification of potential anticholinesterases [7]. The majority of plant-derived compounds with anticholinesterase activity can be

Sathya Sambavathas, Faculty of Allied Health Sciences, University of Jaffna, Jaffna, Sri Lanka
Nilupa R. Amarasinghe, Faculty of Allied Health Sciences, University of Peradeniya, Peradeniya, Sri Lanka
Lalith Jayasinghe, National Institute of Fundamental Studies, Hantana Road, Kandy, Sri Lanka, e-mail: lalith.ja@nifs.ac.lk and ulbj2003@yahoo.com
Yoshinori Fujimoto, National Institute of Fundamental Studies, Hantana Road, Kandy, Sri Lanka; School of Agriculture, Meiji University, Kawasaki, Kanagawa 214-8571, Japan

https://doi.org/10.1515/9783110595949-007

categorized into alkaloids, terpenes, sulfur compounds, and phenolic compounds including flavonoids, benzenoids, stilbenes, lignans, oxygen heterocycles, etc [2].

Simple and rapid screening methods are essential for finding novel therapeutic agents with AChE inhibitory activity. Popular screening methods include UV-visible spectrometric assays, fluorimetric assays, diffractometric assays, mass spectrometric assays, histochemical localization of AChE, colorimetric sticks or strips, radiometric assays, TLC bioautography assays, biosensors and chip techniques [1]. Among these, common methods are UV spectrometric assays and TLC bioautography method based on Ellman's [8] and Marston's [9] methods.

7.2 Spices with AChE inhibitory activity

Spices are a vital component in an everyday meal among South Asians and are culturally linked with the lifestyle. Recent studies have shown that spices provide much more health benefits than just a flavor in food [10]. Research findings over the past 10 years have indicated that phytochemicals derived from various spices such as *Cinnamomum zeylanicum*, *Coriandrum sativum*, *Curcuma longa*, *Garcinia cambogia*, *Myristica fragrans*, *Piper nigrum*, *Syzygium aromaticum*, and *Tamarindus indica* slow down or delay the onset of neurological diseases via multiple mechanisms [6].

a) *Cinnamomum zeylanicum* Blume (cinnamon)

Ceylon cinnamon or true cinnamon, is the dried bark of *Cinnamomum zeylanicum*, belongs to the family Lauraceae. *C. zeylanicum* is used both in the food industry and in indigenous medicine. *C. zeylanicum* is the most important spice, which was used since ancient times for many purposes such as medicine, spice, perfumery material, and soft drink. Sri Lanka is the world's largest producer and exporter of the best quality cinnamon to the world [11]. It possesses many medicinal properties such as sugar control, antioxidant, anti-inflammatory, and antimicrobial activity [12]. Traditionally it was also used for diabetes in Ayurveda and Chinese medicine. *C. zeylanicum* is used to treat stomachic and carminative for gastrointestinal complaints as well as other ailments, including toothache [13].

Water and ethanol (EtOH) extracts of bark of *C. zeylanicum* showed anticholinesterase activity of 46.84 ± 0.003% and 40.83 ± 0.005%, respectively, at 100 µg/mL [14]. Eugenol (Figure 7.1; (1)), one of the major components of the bark of *C. zeylanicum* showed anticholinesterase activity of 0.48 ± 0.16 mg/mL and the positive control galantamine 0.38 ± 0.16 µg/mL [15].

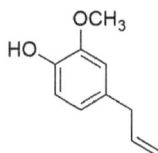

1   **Figure 7.1:** Anticholinesterase active compound of *C. zeylanicum*.

b) *Coriandrum sativum* L. (coriander)

Coriandrum sativum belongs to the family of Apiaceae and is native to Western Asia, Eastern Mediterranean region and Europe. Leaves of *C. sativum* have a distinctive aroma. Both leaves and dried fruits are used as spice [16]. *C. sativum* is used in traditional Indian medicine to treat digestive, respiratory, and urinary system disorders. In the European traditional medical system, the fruits of *C. sativum* are used to strengthen memory [17]. *C. sativum* is reported to have interesting biological activities including antioxidant, anti-mutagenic, antihelmintic, sedative-hypnotic, anticonvulsant, antimicrobial, diuretic, cholesterol-lowering, hypolipidemic, hypoglycemic, antifeedant, anticancer, anxiolytic, hepatoprotective, cardioprotective, antiprotozoal, antiulcer, post-coital anti-fertility and heavy metal detoxification activities [18].

Methanol (MeOH) extract of Sri Lankan *C. sativum* seeds did not exhibit any AChE inhibitory activity in the assay, whereas *n*-hexane and dichloromethane (DCM) extracts showed low activity with IC_{50} values greater than 100 µg/mL. Ethyl acetate (EtOAc) extract showed the highest activity (IC_{50}, 63.51 ± 0.08 µg/mL) among the four extracts [19]. Previous studies have shown leaf MeOH extract to have AChE inhibitory activity of $36.25 \pm 5.3\%$ at 0.1 mg/mL where positive control physostigmine showed IC_{50} of 0.075 ± 0.003 µg/mL [20]. Further, fresh leaves of *C. sativum* showed a dose-dependent improvement in memory scores of young as well as aged rats [21].

c) *Curcuma longa* L. (turmeric)

Curcuma longa belongs to the family of ginger, Zingiberaceae, and it is indigenous to the Indian subcontinent [22]. Turmeric is widely used in the indigenous system of medicine and Ayurvedic systems for centuries to treat various ailments including inflammation and diseases such as biliary disorders, anorexia, cough, diabetic wounds, hepatic disorders, rheumatism, and sinusitis [18]. Turmeric is one of the most commonly used herbal medicine with various activities including antioxidant, apoptotic, antidepressant, antifungal, antiplatelet, antispasmodic, antiarthritic, hypoglycemic, hypotensive, antibacterial, leishmanicidal, antigenotoxicity, cardioprotective, neuroprotective, wound healing, cytoprotective by induction of heat shock protein [23]. Curcuminoid (Figure 7.2; curcumin (**2**), demethoxycurcumin (DMC) (**3**) and bisdemethoxycurcumin (BMC) (**4**)) is the important class of compounds that is responsible for most of the bioactivities mentioned above [24].

Kalaycıoğlu and co-workers reported BMC (**4**) to have the highest anticholines-terase activity (IC_{50} 2.14 ± 0.78 µM) while the positive control galantamine had IC_{50} 2.41 ± 0.12 µM. The lowest activity (IC_{50} 51.8 ± 0.6 µM) was found in curcumin (**2**) [25]. *In vitro* studies indicated that curcumin (**2**) has anti-amyloidogenic, antioxida-tive, anticholinesterase, β-secretase and anti-inflammatory properties while *in vivo* studies resulted in inhibition of amyloid beta ($A\beta$) deposition, $A\beta$ oligomerization, and tau phosphorylation in the brains of AD animal models and improvements in behavioral impairment in animal models with the potential to prevent AD [24]. Though curcumin (**2**) is one of the promising candidates for AD, low bioavailability after oral administration limits its use. Many studies were conducted with intrana-sal route of administration for the central nervous system [26]. A study conducted in combination therapy of curcumin and donepezil supports the concept that the combination strategy might be an alternative therapy in the management/preven-tion of neurological disorders [27]. Several preclinical studies have resulted in bene-ficial effects of curcumin in AD, although the number of human studies is limited. According to these results, curcumin may stabilize or prevent cognitive decline [28].

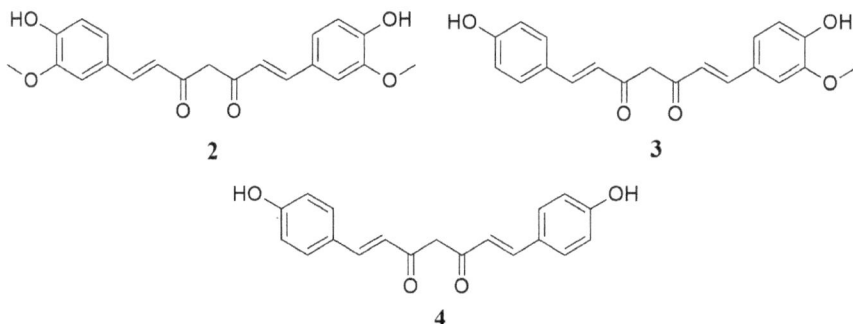

Figure 7.2: Structures of anticholinesterase active curcuminoids.

d) *Garcinia cambogia* Desr. (garcinia)

Garcinia cambogia (formally *Garcinia gummi-gutta*) belongs to the family Clusiaceae and is grown in South Asian countries. The herbal preparations made using *G. cambogia* rinds are used to treat inflammatory ailments, rheumatic pains, and bowel complaints. The fruit is considered to be anthelmintic and cardiotonic. The juice (sherbet) made from the rind is used for piles, hemorrhoids, colic problems, ulcers, inflammations, treat sores, dermatitis, diarrhea, dysentery, ear infections, to facilitate digestion and to pre-vent hyper perspiration [29]. Biological activities of *G. cambogia* include appetite-suppressant, antiobesity, hypolipidemic, antidiabetic, anti-inflammatory, antioxidant, hepatoprotective, anticancer, antiulcer, anticholinesterase, antimicrobial, anthelmin-tic, and diuretic activities, and effects on fertility such as increasing sperm count and lowering the level of testis meiosis-activating sterol responsible for spermatogenesis [30]. *n*-Hexane and DCM extracts from fruits of Sri Lankan *G. cambogia* showed high activity against AChE with IC_{50} values of 42.74 ± 0.10 µg/mL and 61.44 ± 0.08 µg/mL [19]. Another study indicated aqueous extract of fruit rind of *G. cambogia* (67.3% at 1 mg/mL concentration) showed a dose-dependent inhibition of AChE extracted from heart homogenate of male rat of Wistar strain. The positive control neostigmine inhibited AChE to an extent of 92% at 1 mg/mL [31]. Garcinol (Figure 7.3; (**5**)), a major compound isolated from *G. cambogia* fruit, exerted anticholinesterase activity of IC_{50} of 0.66 ± 0.02 µM, which is very close to the positive control galantamine (IC_{50} 0.5 ± 0.01 µM) [32]. Liao and co-workers showed that garcinol acts as a neuroprotective agent by inhibiting the expression of inducible nitric oxide synthase and cyclooxygenase-2 in lipopolysaccharide activated macrophages [33].

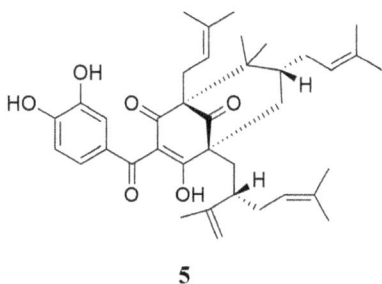

5

Figure 7.3: Structure of anticholinesterase active compound of *G. cambogia*.

e) *Myristica fragrans* Houtt. (nutmeg and mace)

Myristica fragrans is a medium- to a large-sized, aromatic evergreen tree that belongs to the family Myristicaceae. Various parts of this plant are used in the food industry and indigenous medicine. *M. fragrans* seeds are used to support digestion, treat diarrhea, rheumatism, and to improve cognitive activity of patients having AD, as well as to treat mouth sores and insomnia. Essential oil from *M. fragrans* seeds has stress-relieving activity; hence, it is used in aromatherapy [34]. Seeds of *M. fragrans* showed memory-enhancing activity attributed to a variety of properties, including antioxidant, anti-inflammatory, and procholinergic activities [35].

Seeds of *M. fragrans* relieve the symptoms of Alzheimer's disease, as they directly affect AChE activity in the brain [34]. Aqueous alcoholic extract of *M. fragrans* seeds inhibited 50% of AChE activity at concentrations of 100–150 µg/mL in a study that has used AChE obtained from bovine erythrocytes [36] and 50% aqueous methanol extract showed an IC_{50} value of $1,024 \pm 11$ µg/mL [37]. Thirteen compounds were isolated from the MeOH extract of *M. fragrans* seeds, among which 4-[(1*S*)-1-hydroxy-3-(4-hydroxy-3-methoxyphenyl)propyl]-1,3-benzenediol (Figure 7.4; (**6**)) showed the highest anticholinesterase activity [38]. *n*-Hexane extract of *M. fragrans* seeds significantly inhibited AChE activity in the brain of Swiss albino mice [39]. Monomer compounds having an acylphenol moiety; malabaricones A, B, and C (Figure 7.4; (**7–9**)) exhibited significant interaction with AChE, showing IC_{50} values 11.0 ± 2.1 µM, 9.0 ± 1.6 µM, 11.7 ± 2.5 µM, respectively (IC_{50}, 0.11 ± 0.02 µM for positive control tacrine) [40]. *n*-Hexane, DCM, EtOAc and MeOH extracts of dried fruit aril of Sri Lankan *M. fragrans* showed high AChE inhibitory activity with IC_{50} values of 29.03 ± 0.11 µg/mL, 21.37 ± 0.07 µg/mL, 18.29 ± 0.04 µg/mL and 13.44 ± 0.13 µg/mL respectively in modified Ellman's method. Chromatographic separation of the combined EtOAc (87.51% inhibition at 100 µg/mL) and MeOH (96.75% inhibition at 100 µg/mL) extracts yield six compounds, out of which malabaricone C (**9**) showed highest anticholinesterase activity with IC_{50} 2.06 ± 0.04 µg/mL while donepezil showed IC_{50} 0.03 ± 0.00 µg/mL [41].

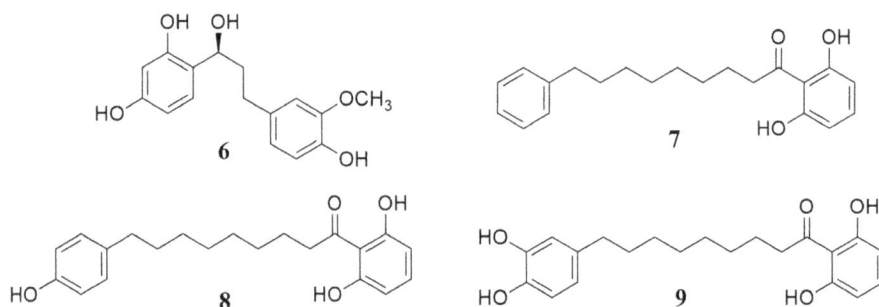

Figure 7.4: Anticholinesterase active compounds of *M. fragrans*.

f) *Piper nigrum* L. (pepper)

Piper nigrum, known as the king of spices, belongs to the family of Piperaceae. *P. nigrum* is a widely used spice around the world and occupies a large percentage of the spice trade. Mature dried berries of the woody perennial evergreen climbing vine are used as the spice. *P. nigrum* is used as a medicinal agent, a preservative and perfumery ingredient [42]. *P. nigrum* is used in a variety of traditional medicinal systems such as Traditional Chinese Medicine, the Indian Ayurvedic system, and folklore medicine of Latin America and West Indies [43]. It is used in folk medicine for stomach disorders, digestive problems, neuralgia, and as a central nervous system depressant [12] *P. nigrum* is used in the treatment of pain relief, chills, rheumatism, flu, muscular aches, colds, exhaustion, and fevers, as a nerve tonic, to increase circulation of blood, increase the flow of saliva, stimulate appetite and to encourage peristalsis [44]. *P. nigrum* shows a diverse array of bioactivities including antihypertensive, anti-asthmatic, cognitive and fertility improvement, antimicrobial, antioxidant, anticancer, anti-inflammatory, hepatoprotective, antidiarrheal, antidepressant, immunomodulatory, anticonvulsant and analgesic activities [42]. Aqueous and EtOH extract of *P. nigrum* seeds showed $52.25 \pm 0.002\%$ and $50.72 \pm 0.002\%$ inhibition respectively at 100 µg/mL, following modified Ellman's method [14]. Seed MeOH extract of *P. nigrum* showed 34% inhibition of AChE at 1,000 µg/mL while galantamine showed an IC_{50} value of 9.4 µg/mL [45]. *n*-Hexane, DCM, EtOH and aqueous extracts of *P. nigrum* seeds showed inhibition of 48.7 ± 3.4%, 42.3 ± 1.3%, 46.8 ± 2.7%, 48.8 ± 2.1% respectively at 500 µg/mL [46]. *n*-Hexane, DCM, EtOAc and MeOH extracts of Sri Lankan *P. nigrum* seeds showed low anticholinesterase activity as their IC_{50} values were higher than 100 µg/mL [19]. Piperine (Figure 7.5; (**10**)) an active alkaloid in *P. nigrum* improves cognitive deficit condition. The mechanism of action of piperine on cognitive improvement is based on the increase of neuron density and anticholinesterase activity in the hippocampus [47].

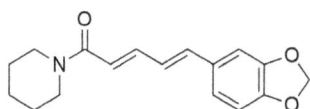

10

Figure 7.5: Anticholinesterase active compounds of *P. nigrum*.

g) *Syzygium aromaticum* (L.) Merr. & L.M. Perry (clove)

Syzygium aromaticum (formally *Eugenia caryophyllus*) is one of the economically important crops in Asian countries, belongs to the family of Myrtaceae. Fully grown, unopened flower buds of this plant are used as a spice all over the world [48]. Whole dried bud or ground powder of *S. aromaticum* is extensively used in Asian cuisine to enhance flavor due to its aroma. Clove oil is obtained by the steam distillation of flower buds, inflorescence branches left after the removal of flower buds. *S. aromaticum* oil is used in the pharmaceutical industry and for perfumery purposes [49]. Dried flower buds of *S. aromaticum* are traditionally used as a carminative to treat hypochlorhydria by increasing hydrochloric acid in the stomach and increasing peristalsis [50]. Ayurvedic uses include antispasmodic, antiemetic, stimulant, dyspepsia, gastric irritation [51]. Several biological activities such as anticancer, antidiabetic, anti-inflammatory, antioxidant, antiulcerogenic, antithrombotic, antifungal, antiviral, antiseptic, antimutagenic, and antiparasitic activities have been reported [52].

A study conducted on aqueous and EtOH extracts of flower bud showed 50.45 ± 0.003% and 59.09 ± 0.006% inhibition respectively at 100 µg/mL [14]. Cold and hot extracts of flower bud showed 90% and 73.7% of inhibition at 100 µg/mL for *in vitro* AChE assay [53]. The essential oil fraction of *S. aromaticum* had 31.94 ± 1.91% inhibition at 120 µg/mL [54]. In another study, MeOH extract showed 47.0% inhibition at 1,000 µg/mL [45]. A study revealed that *S. aromaticum* oil, MeOH extract of the dried flower bud of *S. aromaticum* and eugenol has potential anticholinesterase activity. Essential oil of the clove bud showed high anticholinesterase activity (IC_{50} 49.73 ± 1.33 µg/mL), compared to MeOH extract (IC_{50} 61.5 ± 1.88 µg/mL) while standard compound eugenol (**1**) showed IC_{50} of 49.73 ± 1.33 µg/mL and positive control galantamine had IC_{50} value of 10.14 ± 0.71 µg/mL. Further TLC bioautography method following Ellman's protocol confirmed the potential anticholinesterase activity of the

MeOH extract of *S. aromaticum* [55]. A study on the leaves (*n*-hexane, EtOAc and MeOH extracts) and bud (MeOH extract) of *S. aromaticum*, revealed that MeOH extract of leaves showed slightly higher activity (IC_{50} 42.10 ± 0.41 μg/mL) than MeOH extract of bud (IC_{50} 45.25 ± 0.07 μg/mL), EtOAc extract of leaves (IC_{50} 55.90 ± 3.82 μg/mL) and *n*-hexane (IC_{50} 62.5 ± 16.62 μg/mL) extract of leaves [56]. *n*-Hexane, DCM, EtOAc and MeOH extracts of Sri Lankan *S. aromaticum* showed IC_{50} value higher than 100 μg/mL with 36.19%, 34.95%, 42.45%, 31.94% inhibitions at 100 μg/mL [19].

h) *Tamarindus indica* L. (tamarind)

Tamarindus indica is a large evergreen tree belonging to the family of Fabaceae and native to African and Asian countries. The fruit pulp of *T. indica* is used as a flavoring agent in food, beverages and lozenges due to the taste of tartaric acid and reducing sugars. Fruits, seeds, leaves, bark, and flowers of *T. indica* are used for various purposes including culinary and medicine [57, 58] Tamarind products, leaves, fruits, and seeds have been extensively used in Indian Ayurvedic medicine and traditional African medicine. Tamarind seeds are used in Cambodia and India, in powdered form, to treat boils and dysentery. Boiled and pounded seeds are reported to treat ulcers and bladder stones and powdered seed husks are used to treat diabetes [58].

DCM extract of *T. indica* fruit pulp showed an inhibition of 26.64% at 100 μg/mL, while MeOH extract of *T. indica* seeds showed significant AChE inhibitory activity with IC_{50} 15.88 ± 0.01 μg/mL [19]. Previous studies indicated that MeOH extract from bark had more potent AChE inhibitory activity (IC_{50} of 268.09 μg/mL) than seed MeOH extract (IC_{50} 287.15 μg/mL). Donepezil inhibited the AChE activity with almost 92.57% inhibition at 100 μg/mL [59].

7.3 Other culinary herbs with AChE inhibitory activity

a) *Allium cepa* L. (onion)

Allium cepa is one of the oldest cultivated plants, belongs to the family of Amaryllidaceae which is utilized worldwide as a culinary herb. The bulb of *A. cepa* has a characteristic flavor and odor due to sulfur compounds present. Apart from the culinary virtues, it is widely used in traditional medicine in a wide variety of internal and external preparations to treat various ailments, including digestive problems, skin diseases, metabolic diseases, insect bites, and others. Experimental studies have proven the pharmacological properties of *A. cepa* including anticancer, antihypertensive, antidiabetic, antimicrobial, analgesic, antioxidant, immunomodulatory, anti-inflammatory, antioxidant, and neuroprotective activities. *A. cepa* is a rich source of phenolic compounds, especially quercetin, anthocyanins, kaempferol and their glycosides, phenolic acids, thiosulfinates, vitamins, and minerals [60]. Most of the reported activities are due to the bioactive polyphenolic compounds present in onion.

Anticholinesterase activity of *Allium cepa* has been investigated using aqueous methanol extract following Ellman's method. Significant AChE inhibitory activity was observed with IC_{50} of 51.78 ± 1.05 µg/mL. Hence onion is considered an excellent candidate for developing as a drug for the management of AD [61]. The same research group has further investigated anticholinesterase activities of standardized EtOAc fraction obtained from aqueous methanol extract. Activity guided fractionation followed by the evaluation of AChE inhibitory potential revealed that the most active fraction had an IC_{50} value of 18.33 ± 1.36 µg/mL, whereas positive control donepezil had an IC_{50} value of 7.06 ± 0.13 µg/mL. Quercetin (Figure 7.6; **(11)**) and quercetin 4' *O*-glucoside (spiraeoside) content in the active fraction was determined using a validated thin layer chromatography densitometric method. The active fraction was further examined using a streptozotocin (STZ) induced mice model of Alzheimer's disease. AChE inhibitory activity and oxidative stress markers were assessed in the brain homogenates of mice. This study indicated quercetin significantly improves the spatial learning and memory impairments caused by STZ. In addition, quercetin was found to act as a good AChE inhibitor in the cerebral cortex and hippocampus [62].

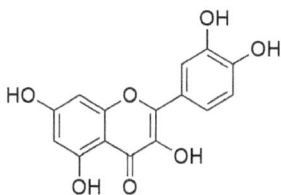

11

Figure 7.6: Anticholinesterase active compounds of *A. cepa*.

b) *Allium sativum* L. (garlic)

Allium sativum is one of the most researched herbal medicines and belongs to the family Amaryllidaceae. It is frequently used as a food and a condiment worldwide. In traditional medicine, garlic is used to treat infections, heart disease, diabetes, diarrhea, rheumatism, and many other disorders. The health benefits of *A. sativum* have been scientifically proven. Scientific and clinical studies have demonstrated that garlic possesses cardioprotective, antihypertensive, anticarcinogenic, immunostimulant, antibacterial, antilipidemic, and hypoglycemic properties [63].

The majority of the compounds reported from *A. sativum* are sulfur compounds including ajoenes, thiosulfinates, vinyldithiins, and various sulfides. Studies have shown that the characteristic pungent odor and many medicinal properties are due to the sulfur compounds. Alliin (*S*-allyl cysteine sulfoxide) (Figure 7.7; **(12)**) is the main sulfur compound, which is transformed to allicin (Figure 7.7; **(13)**) by allinase enzyme released during injuring the garlic bulb through cutting or crushing. Allicin (diallyl thiosulfinate) is a lipid-soluble volatile compound and is reported as the most biologically active compound. In addition to organosulfur compounds, garlic contains various enzymes and amino acids and is rich in minerals including selenium [64]. *In vitro* and animal studies have demonstrated that garlic improves cognitive function through multiple mechanisms having modulatory effects on the neurotransmission process, antioxidant and anti-inflammatory activities. Repeated administration of fresh *A. sativum* was found to increase memory retention in rats [65]. Allicin has many pharmacological activities with good therapeutic potential for many neurological disorders through multiple mechanisms. The ability of allicin to inhibit AChE and BChE was investigated by *in vitro* studies. Allicin strongly inhibited AChE activity while BChE activity was weakly inhibited in a concentration-dependent manner (IC$_{50}$ 61.62 µM for AChE and 308.12 µM for BChE). This finding supported the therapeutic activity of garlic against cognitive decline [66].

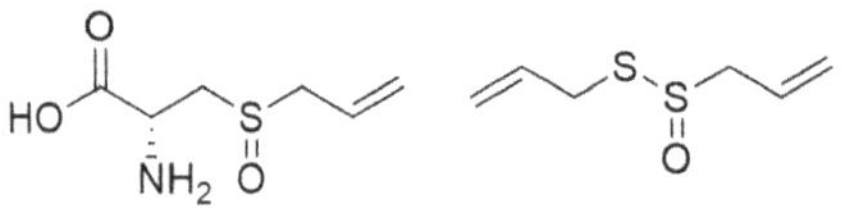

Figure 7.7: Anticholinesterase active compounds of *A. sativum*.

c) *Murraya koenigii* (L.) Spreng. (curry leaves)

Murraya koenigii is commonly known as curry leaves and belongs to the family Rutaceae. It is a popular condiment among Asians. Both fresh leaves and dried powder are added to enhance the flavor and aroma in Asian curries. *M. koenigii* is widely used as a folk medicine to treat stomach aches, rheumatism, influenza, traumatic injury, dysentery, and wounds. Scientific studies have shown that *M. koenigii* possesses hypoglycemic, antidiabetic, hepatoprotective, antibacterial, antioxidant and many other activities. The phytochemical evaluation revealed the presence of a range of carbazole alkaloids, essential oil, and carotenoids [67].

Activity-guided fractionation of petroleum ether extract of *M. koenigii* leaves afforded a carbazole alkaloid, mahanimbine (Figure 7.8; (**14**)) [(3S)-3,11-dihydro-3,5-dimethyl-3-(4-methylpent-3-enyl)pyrano[3,2-a]carbazole]. AChE inhibition of mahanimbine was evaluated following Ellman's method. Mahanimbine inhibited AChE activity in a dose-dependent manner with an IC$_{50}$ value of 30 ± 90 µg/mL. The positive control galantamine showed an IC$_{50}$ value of 6 ± 1 µg/mL. This study revealed *M. koenigii* processes promising compounds that can be further developed as potential therapeutic agents against AD [68].

Another study was conducted to evaluate the effect of *M. koenigii* alkaloidal extract on cognitive functions and brain cholinesterase activity of mice. When leaf alkaloidal extract was orally administered in three doses (10, 20, and 30 mg/kg) to young

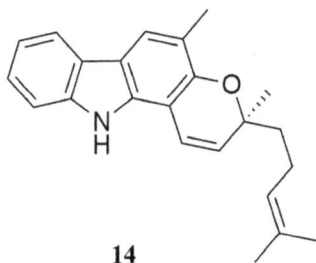

14

Figure 7.8: Anticholinesterase active compounds of *M. koenigii.*

and aged mice, significant improvement in memory scores of both mice was observed for 20 and 30 mg/kg doses. Further, the brain cholinesterase activity was significantly reduced, thus suggesting the potential therapeutic activity of the alkaloidal extract in managing the symptoms of AD and dementia. This study revealed the alkaloids extract of *M. koenigii* leaves increased the brain acetylcholine levels and thereby improved the cognitive function of both young and aged mice. *In vitro* evaluation of the same extract showed BChE inhibition through a non-competitive mechanism. This study further supported the therapeutic potential of *M. koenigii* for the management of Alzheimer patients [69].

d) *Zingiber officinale* Roscoe (ginger)

Zingiber officinale belongs to the family Zingiberaceae and originated in East Asia. The rhizome of this plant is popular around the world due to its pungency and typical aroma. *Z. officinale* is one of the most famous medicinal herbs in the Indian Ayurvedic system and Traditional Chinese Medicine for centuries. Oleoresin of *Z. officinale* is used in food, beverages, soft drinks, and in herbal drugs/products as bioavailability enhancer in anti-inflammatory herbal products. The rhizome of *Z. officinale* is used to treat a wide range of ailments, including the common cold, fever, sore throat, pain, rheumatism, bronchitis as well as carminative for gastrointestinal disorders, nausea and vomiting. In addition, it is used to treat toothache, gingivitis, bronchitis, hypertension, dementia, helminthiasis, constipation, asthmatic respiratory disorders, antispasmodic, expectorant, peripheral circulatory stimulant and astringent [70]. *Z. officinale* possess a wide range of biological activities such as cardioprotective, anticonvulsant, anxiolytic, antiemetic, antidiabetic, hypolipidemic, anti-inflammatory, antithrombotic, antiobesity, antioxidant, antitumor, anti-atherosclerotic, radioprotective, hypotensive, antiulcer, hepatoprotective, etc.

A recent study has shown that a standardized extract of the dry rhizomes of *Z. officinale* affects the initiation and development of neurodegeneration by inhibiting messenger ribonucleic acid (mRNA) expression of pro-inflammatory mediators and amyloid β-induced inflammatory mediators with good potential in the prevention and treatment of AD [70–72]. A study reported negative inhibition for MeOH extract of *Z. officinale* rhizobium [45] while in another study IC$_{50}$ of 41 ± 1.2 µg/mL was observed for MeOH extract of Z. officinale rhizobium and 0.075 ± 0.003 µg/mL for physostigmine [73]. 60% EtOH extract of Z. officinale showed an inhibition of $48.04 \pm 0.06\%$ at 100 µg/mL while galantamine showed $93.44 \pm 2.21\%$ of inhibition at the same concentration [74]. Zerumbone (Figure 7.9; (15)) present in the Z. officinale exhibited AChE inhibitory activity through TLC bioautographic method following Ellman's method [75]. In a comparative molecular docking approach using AutoDock, it was proposed that gingerenone A (Figure 7.9; (16)) binds to the active site of AChE and appears to interact with AChE conferring minimum binding energy among the docked compounds which facilitate the AChE inhibitory activity [76].

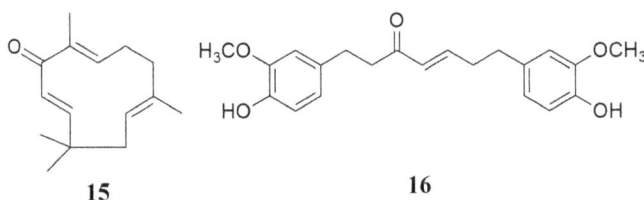

15   16

Figure 7.9: Anticholinesterase active compounds of Z. officinale.

7.4 Conclusion

Selected examples given in this brief chapter indicate the therapeutic potential of spices and other culinary plants used in South Asian cuisine. Most of the investigations were conducted using *in vitro* testing using enzyme assays or using mouse models. However, further investigations, including bioavailability studies, are required to confirm the effectiveness of the active compounds or extracts.

Abbreviations

AChE	Acetylcholinesterase
AD	Alzheimer's disease
BChE	Butyrylcholinesterase
DCM	Dichloromethane
EtOAc	Ethyl acetate
MeOH	Methanol

References

[1]   Miao Y, He N, Zhu JJ. History and new developments of assays for cholinesterase activity and inhibition. Chem Rev 2010, 110(9), 5216–5234.

[2]   Filho JMB, Medeiros KCP, Diniz MDFFM, Batista LM, Athayde-Filho PF, Silva MS, Da Cunha EVL, Almeida JRGS, Quintans-Júnior LJ. Natural products inhibitors of the enzyme acetylcholinesterase. Revisão Rev Bras Farmacogn 2006, 16(2), 258–285.

[3]   Mukherjee PK, Kumar V, Mal M, Houghton PJ. Acetylcholinesterase inhibitors from plants. Phytomedicine 2007, 14(4), 289–300.

[4]   Raj SR, Black BK, Biaggioni I, Harris PA, Robertson D. Acetylcholinesterase inhibition improves tachycardia in postural tachycardia syndrome. Circulation 2005, 111(21), 2734–2740.

[5]   World Health Organization, Global action plan on the public health response to dementia. 2017–2025. Geneva World Health Organization, 2017, 52.

[6]   Ong WY, Farooqui T, Ho CF, Ng Y-K, Farooqui AA. Use of phytochemicals against neuroinflammation. Farooqui T, Farooqui AA, eds, Neuroprotective effects of phytochemicals in neurological disorders. 1st, New Jersey, USA, Wiley Blackwell, 2017, 1–41.

[7]   Suganthy N, Pandian SK, Devi KP. Cholinesterase inhibitors from plants: Possible treatment strategy for neurological disorders–A review. Int J Biomed Pharm Sci 2009, 3(Special Issue 1), 87–103.

[8]   Ellman GL, Courtney KD, Andres V Jr, Featherstone RM. A new and rapid colorimetric of acetylcholinesterase determination of acetylcholinesterase activity. Biochem Pharmacol 1961, 7, 88–95.

[9]   Marston A, Kissling J, Hostettmann K. A rapid TLC bioautographic method for the detection of acetylcholinesterase and butyrylcholinesterase inhibitors in plants. Phytochem Anal 2002, 13(1), 51–54.

[10] Rathore SS, Saxena SN, Singh B. Potential health benefits of major seed spices. Int J Seed Spices 2013, 3(2), 1–12.

[11] Sri Lanka Export Development Board. Ceylon cinnamon as spice. A Digit. Mark. Solut. by eBEYONDS, 2020. (Accessed May 16, 2020, at https://www.srilankabusiness.com/spices/ceylon-cinnamon-as-spice.html.)

[12] Ranjan N, Kumari M. Acetylcholinesterase inhibition by medicinal plants: A Review. Ann Plant Sci 2017, 6, 1640–1644.

[13] Bandara T, Uluwaduge I, Jansz ER. Bioactivity of cinnamon with special emphasis on diabetes mellitus: A review. Int J Food Sci Nutr 2012, 63(3), 380–386.

[14] Kumar S, Brijesh L, Dixit S. Screening of traditional Indian spices for inhibitory activity of acetylcholinesterase and butyrylcholinesterase enzymes. Int J Pharma Bio Sci 2012, 3(1), 59–65.

[15] Dohi S, Terasaki M, Makino M. Acetylcholinesterase inhibitory activity and chemical composition of commercial essential oils. J Agric Food Chem 2009, 57(10), 4313–4318.

[16] Mandal S, Mandal M. Coriander (*Coriandrum sativum* L.) essential oil: Chemistry and biological activity. Asian Pac J Trop Biomed 2015, 5(6), 421–428.

[17] Lobbens ES, Vissing KJ, Jorgensen L, van de Weert M, Jäger AK. Screening of plants used in the European traditional medicine to treat memory disorders for acetylcholinesterase inhibitory activity and anti-amyloidogenic activity. J Ethnopharmacol 2017, 200, 66–73.

[18] Phukan P, Bawari M, Sengupta M. Promising neuroprotective plants from North-East India. Int J Pharm Pharm Sci 2015, 7(3), 28–39.

[19] Sathya S Investigation of acetylcholinesterase inhibitory activity of selected Sri Lankan grown spices as potential therapeutic agents for Alzheimer's disease. M.Phil. Thesis, Postgraduate Institute of Science, University of Peradeniya, Sri Lanka, 2019.

[20] Mathew M, Subramanian S. *In vitro* Screening for anti-cholinesterase and antioxidant activity of methanolic extracts of Ayurvedic medicinal plants used for cognitive disorders. PLoS One 2014, 9(1), e86804. 10.1371/journal.pone.0086804.

[21] Mani V, Parle M. Memory-enhancing activity of *Coriandrum sativum* in rats. Pharmacologyonline 2009, 2, 827–839.

[22] Madhusankha G, Thilakarathan R, Liyanage T, Navaratne SB. Compositional analysis of turmeric types cultivated in Sri Lanka and India. Int J Herb Med 2019, 7(1), 35–38.

[23] Bhat SV, Amin T, Nazir S. Biological activities of turmeric (*Curcuma longa* Linn.) - An overview. BMR Microbiol 2015, 1(1), 1–5.

[24] Hamaguchi T, Ono K, Yamada M. Review: Curcumin and Alzheimer's disease. CNS Neurosci Ther 2010, 16(5), 285–297.

[25] Kalaycıoğlu Z, Gazioğlu I, Erim FB. Comparison of antioxidant, anticholinesterase, and antidiabetic activities of three curcuminoids isolated from *Curcuma longa* L. Nat Prod Res 2017, 31(24), 2914–2917.

[26] Dimauro TM Intranasally administering curcumin to the brain to treat Alzheimer's disease. 2008. (Accessed May 7, 2020, at https://patents.google.com/patent/US20080075671A1/en.)

[27] Akinyemi AJ, Oboh G, Oyeleye SI, Ogunsuyi O. Anti-amnestic effect of curcumin in combination with donepezil, an anticholinesterase drug: Involvement of cholinergic system. Neurotox Res 2017, 31(4), 560–569.

[28] Voulgaropoulou SD, Van Amelsvoort TAMJ, Prickaerts J, Vingerhoets C. The effect of curcumin on cognition in Alzheimer's disease and healthy aging: A systematic review of pre-clinical and clinical studies. Brain Res 2019, 1725. 46476 10.1016/j.brainres.2019.146476. Epub 2019.

[29] Rasha H, Salha A, Thanai A, Zahar A. The biological importance of *Garcinia cambogia*: A review. J Nutr Food Sci 2012, s5, 004, 10.4172/2155-9600.S5-004.

[30] Semwal RB, Semwal DK, Vermaak I, Viljoen A. A comprehensive scientific overview of *Garcinia cambogia*. Fitoterapia 2015, 102, 134–148.

[31] Subhashini N, Nagarajan G, Kavimani S. *In vitro* antioxidant and anticholinesterase activities of *Garcinia cambogia*. Int J Pharm Pharm Sci 2011, 3(3), 129–132.

[32] Lenta BN, Vonthron-Sénécheau C, Weniger B, Devkota KP, Ngoupayo J, Kaiser M, Naz Q, Choudhary MI, Tsamo E, Sewald N. Leishmanicidal and cholinesterase inhibiting activities of phenolic compounds from *Allanblackia monticola* and *Symphonia globulifera*. Molecules 2007, 12(8), 1548–1557.

[33] Liao CH, Sang S, Liang YC, Ho CT, Lin JK. Suppression of inducible nitric oxide synthase and cyclooxygenase-2 in downregulating nuclear factor-kappa β pathway by garcinol. Mol Carcinog 2004, 41(3), 140–149.

[34] Gupta AD, Rajpurohit D. Antioxidant and antimicrobial activity of nutmeg (*Myristica fragrans*). Victor R, Watson RR, Patel VB, eds, Nuts and seeds in health and disease prevention. London, UK, Acadamic Press, 2011, 831–839.

[35] Jaiswal P, Kumar P, Singh VK, Singh DK. Biological effects of *Myristica fragrans*. Annu Rev Biomed Sci 2009, 11, 21–29.

[36] Jaiswal P, Kumar P, Singh VK, Singh DK. Enzyme inhibition by molluscicidal components of *Myristica fragrans* Houtt. in the nervous tissue of snail *Lymnaea acuminata*. Enzyme Res 2010, 2010, 478746, 10.4061/2010/478746.

[37] Jazayeri SB, Amanlou A, Ghanadian N, Pasalar P, Amanlou M. A preliminary investigation of anticholinesterase activity of some Iranian medicinal plants commonly used in traditional medicine. DARU J Pharm Sci 2014, 22(1), 17. 10.1186/2008-2231-22-17.

[38] Cuong TD, Hung TM, Han HY, Roh HS, Seok JH, Lee JK, Jeong JY, Choi JS, Kim JA, Min BS. Potent acetylcholinesterase inhibitory compounds from *Myristica fragrans*. Nat Prod Commun 2014, 9, 499–502.

[39] Dhingra D, Parle M, Kulkarni SK. Comparative brain cholinesterase-inhibiting activity of *Glycyrrhiza glabra, Myristica fragrans*, ascorbic acid, and metrifonate in mice. J Med Food 2006, 9(2), 281–283.

[40] Maia A, Schmitz-Afonso I, Martin MT, Awang K, Laprévote O, Guéritte F, Litaudon M. Acylphenols from *Myristica crassa* as new acetylcholinesterase inhibitors. Planta Med 2008, 74(12), 1457–1462.

[41] Sathya S, Amarasinghe NR, Jayasinghe L, Araya H, Fujimoto Y. Enzyme inhibitors from the aril of *Myristica fragrans*. S Afr J Bot 2020, 130, 172–176.

[42] Damanhouri ZA, Ahmad A. A Review on therapeutic potential of *Piper nigrum* L. (black pepper): The king of spices. Med Aromat Plants 2014, 3(3), 161.

[43] Ghosh R, Darin K, Nath P, Deb P. An overview of various piper species for their biological activities. Int J Pharma Res Rev 2014, 3(1), 67–75.

[44] Meghwal M, Goswami TK. *Piper nigrum* and piperine: An update. Phyther Res 2013, 27(8), 1121–1130.

[45] Ali SK, Hamed AR, Soltan MM, Hegazy UM, Elgorashi EE, El-Garf IA, Hussein AA. In *vitro* evaluation of selected Egyptian traditional herbal medicines for treatment of Alzheimer disease. BMC Complement Altern Med 2013, 13, 121, 10.1186/1472-6882-13-121.

[46] Gupta M, Sharma C, Meena P, Khatri M. Investigating the free radical scavenging and acetylcholinesterase inhibition activities of *Elletaria cardamomum, Piper nigrum* and *Syzygium aromaticum*. Int J Pharm Sci Res 2017, 8(7), 3180–3186.

[47] Chonpathompikunlert P, Wattanathorn J, Muchimapura S. Piperine, the main alkaloid of Thai black pepper, protects against neurodegeneration and cognitive impairment in animal model of cognitive deficit like condition of Alzheimer's disease. Food Chem Toxicol 2010, 48(3), 798–802.

[48] Spice council Sri Lanka. Sri Lankan Spices. Spice Counc. Sri Lanka. 2010, (Accessed May 15, 2020, at http://www.srilankanspices.com/sl_spices_cardamom.html.)

[49] Mittal M, Gupta N, Parashar P, Mehra V, Khatri M. Phytochemical evaluation and pharmacological activity of *Syzygium aromaticum*: A comprehensive review. Int J Pharm Pharm Sci 2014, 6(8), 67–72.

[50] Milind P, Clove: DK. A champion spice. Int J Res Ayurveda Pharm 2011, 2(1), 47–54.

[51] Jacquet ADR, Subedi R, Ghimire SK, Rochet JC. Nepalese traditional medicine and symptoms related to Parkinsons disease and other disorders: Patterns of the usage of plant resources along the Himalayan altitudinal range. J Ethnopharmacol 2014, 153(1), 178–189.

[52] Mahalwal VS. Volatile constituents and biological activities of dried unripe flower buds of *Syzygium aromaticum* Linn. Syn. *Eugenia caryophyllus* (Spreng.). Bullock & S. G. Harrison Int J Green Pharm 2017, 11(4), 8–12.

[53] Prabha M, Anusha TS. Esterase's properties in commonly used Indian spices for Alzheimer's disease model. J Biochem Technol 2015, 6(1), 875–882.

[54] Phrompittayarat W, Hongratanaworakit T, Sarin Tadtong K, Sareedenchai V, Ingkaninan K. Survey of Aacetylcholinesterase inhibitory activity in essential oils from aromatic plants. Int J Medicinal Aromat Plants 2014, 4(1), 1–5.

[55] Dalai MK, Bhadra S, Chaudhary SK, Bandyopadhyay A, Mukherjee PK. Anti-cholinesterase activity of the standardized extract of *Syzygium aromaticum* L. Pharmacogn Mag 2014, 10 (Suppl 2), S276–282.

[56] Darusman LK, Wulan TW, Alwi F. Acetylcholinesterase inhibition and antioxidant activity of *Syzygium cumini*, *S.aromaticum* and *S.polyanthum* from Indonesia. J Biol Sci 2013, 13(5), 412–416.

[57] De Caluwé E, Halamová K, Van Damme P. *Tamarindus indica*: A review of traditional uses, phytochemistry and pharmacology. Afrika Focus 2010, 23(1), 53–83.

[58] Rao SY, Mathew MK. Tamarind. Peter KV, ed, Handbook of herbs and spices- Volume 1. 1st, Cambridge, UK, Woodhead Publishing Limited, 2001, 287–296.

[59] Biswas K, Azad A, Sultana T, Khan F, Hossain S, Alam S, Chowdhary R, Khatun Y. Assessment of *in vitro* cholinesterase inhibitory and thrombolytic potential of bark and seed extracts of *Tamarindus indica* (L.) relevant to the treatment of Alzheimer's disease and clotting disorders. J Intercult Ethnopharmacol 2017, 6(1), 115–120.

[60] Bystrická J, Musilová J, Vollmannová A, Timoracká M, Kavalcová P. Bioactive components of onion (*Allium cepa* L.) – A Review. Acta Aliment 2013, 42(1), 11–22.

[61] Kaur R, Shri R. Evaluation of two culinary plant species for anticholinesterase, antioxidant and cytotoxic activity. Pharmacogn Commun 2018, 8(1), 25–28.

[62] Kaur R, Randhawa K, Kaur S, Shri R. *Allium cepa* fraction attenuates STZ- induced dementia via cholinesterase inhibition and amelioration of oxidative stress in mice. J Basic Clin Physiol Pharmacol 2020, 31(3), 10.1515/jbcpp-2019-0197.

[63] El-Saber BG, Beshbishy MA, Wasef GL, Elewa Y, Al-Sagan AA, Abd El-Hack ME, Taha AE, Abd-Elhakim MY, Devkota PH. Chemical constituents and pharmacological activities of garlic (*Allium sativum* L.): A review. Nutrients 2020, 12(3), 872. 10.3390/nu12030872.

[64] Bhandari PR. Garlic (*Allium sativum* L.): A review of potential therapeutic applications. Int J Green Pharm 2012, 6(2), 118–129.

[65] Haider S, Naz N, Khaliq S, Perveen T, Haleem DJ. Repeated administration of fresh garlic increases memory retention in rats. J Med Food 2008, 11(4), 675–679.

[66] Kumar S. Dual inhibition of acetylcholinesterase and butyrylcholinesterase enzymes by allicin. Indian J Pharmacol 2015, 47(4), 444–446.

[67] Jain V, Momin M, Laddha K. *Murraya koenigii*: An updated review. Int J Ayurvedic Herb Med 2012, 2(04), 607–627.

[68] Kumar NS, Mukherjee PK, Bhadra S, Saha BP, Pal BC. Acetylcholinesterase inhibitory potential of a carbazole alkaloid, mahanimbine, from *Murraya koenigii*. Phyther Res 2010, 24(4), 629–631.

[69] Mani V, Ramasamy K, Ahmad A, Parle M, Shah SA, Majeed AB. Protective effects of total alkaloidal extract from *Murraya koenigii* leaves on experimentally induced dementia. Food Chem Toxicol 2012, 50(3–4), 1036–1044.

[70] Mbaveng AT, Keute V. Zingiber officinale. Keute V, ed, Medicinal spices and vegetables from africa- therapeutic potential against metabolic, inflammatory, infectious and systemic diseases. 1st, London, UK, Academic Press, 2017, 627–639.

[71] Islam MA, Khandker SS, Alam F, Khalil MI, Kamal MA, Gan SH. Alzheimer's disease and natural products: Future regimens emerging from nature. Curr Top Med Chem 2017, 17(12), 1408–1428.

[72] Grzanna R, Phan P, Polotsky A, Lindmark L, Frondoza CG. Ginger extract inhibits beta-amyloid peptide-induced cytokine and chemokine expression in cultured THP-1 monocytes. J Altern Complement Med 2004, 10(6), 1009–1013.

[73] Mathew M, Subramanian S. *In vitro* evaluation of anti-Alzheimer effects of dry ginger (*Zingiber officinale* Roscoe) extract. Indian J Exp Biol 2014, 52(6), 606–612.

[74] Ali-Shtayeh MS, Jamous RM, Zaitoun SYA, Qasem IB, Ali-Shtayeh MS. *In vitro* screening of acetylcholinesterase inhibitory activity of extracts from Palestinian indigenous flora in relation to the treatment of Alzheimer's disease. Funct Foods Heal Dis 2014, 4(9), 381–400.

[75] Bustamam A, Ibrahim S, Al-Zubairi AS, Met M, Syam MM. Zerumbone: A natural compound with anticholinesterase activity. Am J Pharmacol Toxicol 2008, 3(3), 209–211.

[76] Azam F, Amer AM, Abulifa AR, Elzwawi MM. Ginger components as new leads for the design and development of novel multi-targeted anti-Alzheimer's drugs: A computational investigation. Drug Des Devel Ther 2014, 8, 2045–2059.

Chamika Liyanaarachchi, Mayuri Napagoda, Sanjeeva Witharana, Lalith Jayasinghe

8 Photoprotective potential in medicinal plants

8.1 Introduction

The general opinion among humans is that exposing to the direct sunlight is harmful. The ultraviolet (UV) component of the solar radiation is considered as harmful to the human skin. There are three different categories of solar UV radiation classified according to the wavelengths. Those are UV-A (315–400 nm), UV-B (280–315 nm) and UV-C (100–280 nm) [1, 2]. About 90% of UV radiation comprises UV-A which is known as "Aging rays." UV-B is specially known as "Burning rays" and it is more gene-toxic than UV-A. The solar UV radiation contains 4–5% of UV-B. UV-C is the most dangerous type of UV radiation; however, it cannot cause any harm to the human skin because it is effectively filtered by the atmospheric ozone layer [3].

Exposure to UV radiation could lead to acute and chronic effects on the human skin. The acute effects of exposure are short-lived and reversible. These effects include mainly sun burn (erythema), inflammation and pigment darkening. Apart from these, inflammation is somewhat correlated with chronic effects. The excess amounts of reactive oxygen species (ROS) which are generated due to excessive exposure to the UV radiation have a major impact on UV-induced chronic effects such as photocarcinogenesis and photoageing [2–5]. ROS are generally produced in the body in low concentrations by enzymatic and also non-enzymatic reactions. They mainly involve signal transduction pathways, gene transcription, etc. [6, 7]. Singlet oxygen, peroxyl radicals, superoxide radicals, hydroxyl radicals, and peroxynitrite are some examples for ROS. The harmful effects occur with the generation of ROS beyond the threshold limit by UV irradiation, particularly when it exceeds the elaborate anti-oxidative defense mechanisms in the body.

Photosensitizer molecules such as riboflavin, porphyrins, and quinones are responsible for the generation of ROS in mammalian cells upon the exposure to UV radiation. The absorption of photons/energy by these molecules initiates photosensitization reaction leading to an excited state called the singlet excited state. Thereafter two reactions can occur: either falling back to the ground state by emitting

_____

Chamika Liyanaarachchi, Faculty of Medicine, University of Ruhuna, Galle 80 000, Sri Lanka
Mayuri Napagoda, Faculty of Medicine, University of Ruhuna, Galle 80 000, Sri Lanka,
e-mail: mayurinapagoda@yahoo.com
Sanjeeva Witharana, Faculty of Engineering, University of Moratuwa, Moratuwa 10400, Sri Lanka
Lalith Jayasinghe, National Institute of Fundamental Studies, Kandy 20 000, Sri Lanka

https://doi.org/10.1515/9783110595949-008

heat/fluorescence, or an intersystem crossing to form a triplet excited state in photosensitizer molecule. Once this excited photosensitizer molecule undergoes electron transport and energy transfer, then it may lead to the formation of radical and non-radical ROS that can cause modifications to proteins, lipids and DNA and thereby cellular damage [8, 9]. Similarly, ROS can induce changes in collagen and elastin fibers of the skin, distorting the structural integrity of the skin [10].

The UV Index (UVI) is an indicator of the amount of skin damaging UV-A and UV-B radiation expected to reach the earth surface at the time when the sun is highest in the sky [11]. The UVI raises public alertness about the risks of excessive exposure to UV radiation, thus encouraging people to adopt protective measures and thereby reducing the harmful health effects of sun's exposure [12]. The UV Index value, risk levels, and precautions to be taken are summarized in Table 8.1 [13].

Table 8.1: UV Index value, risk levels, and precautions to be taken.

UV index and risk	Action
0–2: Low	– Wear sunglasses on bright or snow days. – Avoid bright surfaces which can reflect UV radiation and thereby increase exposure.
3–5: Moderate	– Use shade whenever possible during daytime when the sun is highest on the sky. – Wear UV blocking sunglasses and cloths. – Use SPF 30$^+$ sunscreen for every 2 hours when in outside.
6–7: High	– Reduce time in the sun between 11am and 4pm and take full precaution by seeking shade
8–10: very high	– Extra protection required – unprotected skin will be damaged and can burn quickly avoid the sun between 11am and 4pm
11$^+$: Extreme	– Take full precaution – unprotected skin will be damaged and can burn in minutes

The term "photoprotection" refers to the effective protection from the harmful effects of UV radiation exposure. There are various kinds of photoprotective strategies which have been used to avoid UV radiation mediated negative biological effects [5].

The best way to avoid UV radiation exposure is staying under the shade during daytime. But it is not always possible because most people are involved in outdoor occupations and activities. Therefore it is necessary to use suitable preventive measures such as sunscreens to avoid the exposure or to reduce the damage caused by UV radiation.

8.2 Sunscreens

Sunscreen is a product that is applied on the skin to protect it from sun's harmful radiation [14]. Topical sunscreen products contain UV absorbing, reflecting, and scattering active molecules [5]. Topical application of a sunscreen could avoid the penetration of UV rays into the skin. The active molecules on the sunscreens are also called photoprotective agents, and they can either prevent the damage caused by UV radiation or modulate different cellular responses to UV radiation to stop tumor promotion and progression. Active molecules of the sunscreens can be either organic or inorganic ingredients [15, 16].

The efficacy of a sunscreen product is determined by calculating a value called sun protection factor (SPF). The SPF for a sunscreen is defined as the ratio of sun exposure that skin can tolerate before burning or minimal erythema is apparent with and without the sunscreen protection [15]. Usually, sunscreens with a sun protection factor (SPF) value of 15 or greater are highly recommended [5].

Most of the sunscreens are produced by combining inorganic and organic ingredients. The most commonly used ingredients in inorganic sunscreens are titanium dioxide (TiO_2) and zinc oxide (ZnO) which can attenuate both UV-A and UV-B radiations via reflection or scattering. Besides, TiO_2 can also absorb a considerable amount of both UV-A and UV-B radiation [17]. On the other hand, ZnO can absorb, reflect, and scatter UV-A radiation [18]. The organic UV filters are usually aromatic compounds that absorb the energy of a specific portion of the UV radiation spectrum and reemit as longer wavelength light or heat. Besides, the absorbed energy can be used for a photochemical reaction, such as cis-trans or keto-enol photochemical isomerization [15]. Every organic UV filter should have chromophores to absorb UV radiation. These chromophores consist of conjugated π-electron systems known as conjugated double bonds. The organic UV filter with a higher number of conjugated double bonds, can give a larger absorption cross section and, hence, higher absorption [2]. Organic UV-filters can be categorized into three groups as UV-A filters (e.g., benzophenones, anthranilates, dibenzoylmethanes), UV-B filters (e.g., salicylates, cinnamates, camphor derivatives), and broad-spectrum filters (e.g., hexyl benzoate, disodium phenyl dibenzimidazole, and drometrizole trisiloxane) depending on their differences in the absorption of UV radiation [2, 16] However, the broad-spectrum UV filters are very scarce. Therefore, the organic sunscreens are always used in combination of UV-A and UV-B filters. The relatively narrow absorption spectrums of these individual organic sunscreens can be broadened by combining with each other [15].

8.2.1 The safety concerns of currently available sunscreen products

Several studies have revealed that the application of sunscreens that contain synthetic organic molecules as UV filters could exert side effects. For example the synthetic organic UV filter para-amino benzoic acid (PABA) has the potential to inhibit the production of Vitamin D after application on the skin. Oxybenzone which is one of the most widely used synthetic organic UV-A filters in the United States, is identified as a potential endocrine disruptor. Furthermore, the studies with rats illustrated that it could get accumulated in several organs such as the liver, kidney, spleen, and intestine [17]. The TiO_2 in sunscreens may exhibit a photocatalytic effect after exposure to UV radiation, although it has been considered as an inert substance. Here, the excitation of TiO_2 by UV-A or UV-B radiation can lead to the formation of ROS such as hydroxyl radicals. The presence of photoreactive TiO_2 in a sunscreen product is also harmful to the organic UV filters that exist in the same sunscreen product because these organic molecules can be oxidized by the produced radicals. Therefore, TiO_2 is coated with other materials before introduction into the sunscreen product [15]. With the escalating number of reports on the adverse effects related to the synthetic sunscreen agents, the recent development in this field is the usage of herbs comprising natural compounds with UV-absorbing property and thereby to reduce and/or minimize the use of synthetic sunscreen products [19, 20].

8.3 Herbal materials as alternative photoprotective agents

The best known photoprotective molecules that can replace the use of synthetic organic molecules are plant-derived natural ingredients, i.e., plant secondary metabolites. The plant secondary metabolites containing aromatic rings are capable of absorbing UV radiation within the wavelength range of 200–400 nm. Many plants synthesize secondary metabolites in response to the exposure to UV radiation [21]. For example, UV-B radiation has identified as a potential elicitor of the synthesis of betalains and flavonoids in the genus *Alternanthera* [22] while the production of flavonoids was increased in *Nymphoides humboldtiana*, resulting in enhancement of its antioxidant activity [23]. These secondary metabolites protect plants from subsequent UV damage and thereby allowing plants to thrive in nature.

Flavonoids are a class of polyphenolic compounds that are being synthesized by the phenylpropanoid metabolic pathway and have been subjected to photoprotective activity studies more than any other group of compounds. Among those, quercetin (Figure 8.1(1)) and rutin (Figure 8.1(2)) are the most widely studied flavonoids [21]. The oil-in-water emulsions prepared by incorporating these two flavonoids exhibited SPF values comparable to that of the standard substance homosalate, while the photoprotection exert by the two flavonoids in the UV-A range was also a non-negligible [24]. Similarly, the topical application of soybean isoflavone genistein (Figure 8.1(3)) prior to UV-B irradiation was capable of effectively inhibiting the UV-B-induced erythema [25]. Moreover, experimental data revealed that vast array of secondary metabolites; specially polyphenols could exert chemoprevention of skin inflammation and subsequently decrease skin tumors resulting from the exposure to carcinogenic doses of UV radiation [26, 27]. Beside the UV-absorbing potential, these natural ingredients have some other beneficial effects that can add a good value to the sunscreen formulation. These effects are mainly anti-inflammatory, anti-mutagenic, and antioxidant [21, 28]. Therefore, sunscreen formulations of natural origin could become the future of cosmetics.

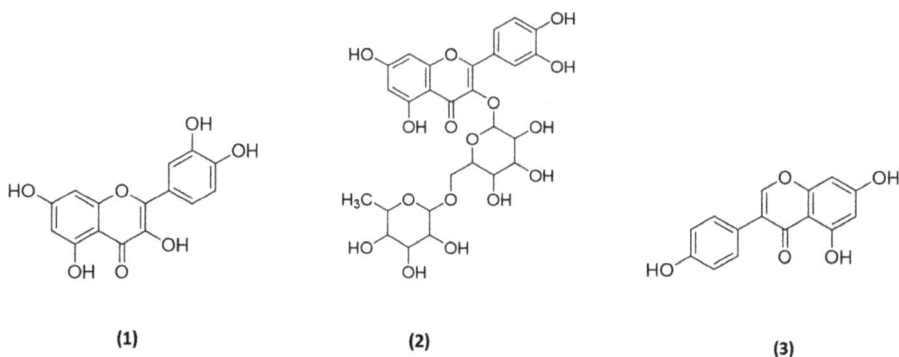

Figure 8.1: Examples of some polyphenolic compounds effective as photoprotectants (1) Quercetin (2) Rutin (3) Genistein.

Melanin is a pigment present in human skin that functions as a major natural defense system against the sun's UV radiation. Therefore an ideal photoprotectant should not inhibit melanin biosynthesis in which tyrosinase acts as a key enzyme. Rosmarinic acid that was extracted from the leaves of rosemary (*Rosmarinus officinalis*) displayed the potential to use as a UV-protector. *R. officinalis* is well known for interesting biological activities, for example, antiviral activity, antibacterial activity as well as anti-inflammatory and antioxidant activities. The rosmarinic acid was capable of reducing the UV-A-induced oxidative stress by scavenging and quenching of ROS which are generated by UV-A exposure. The *in vivo* assays confirmed that oral administration of rosmarinic acid could inhibit cutaneous alterations (photocarcinogenesis) caused by exposition to UV-A. Interestingly, this compound induced the body's own endogenous defense system by increasing the tyrosinase activity and hence stimulating melanin production [29].

The naturally occurring phenolic compounds are useful tools to prevent UV radiation-mediated detrimental effects. The green tea obtained from the plant *Camellia sinensis* possesses phenolic acids and flavonols that inherit antioxidant properties. The topical application of polyphenol fraction isolated from green tea was found to be effective in preventing photocarcinogenesis particularly by decreasing the formation of dimeric DNA photoproducts like cyclobutane pyrimidine dimers [30, 31]. In another experiment, the *in vitro* SPF value of alcoholic dilution of a green tea extract (90% ethyl alcohol) was recorded as 18.10 ± 0.05 confirming the photoprotective potential of green tea polyphenols [32]. A clinical study conducted with 60 female volunteers revealed that the consumption of green tea polyphenol beverage daily for three months had significantly improved skin hydration, density, and elasticity and thereby helped to maintain skin integrity [33]. Earlier, Jeon et al. [34] demonstrated that the regular intake of epigallocatechin gallate (EGCG), an ubiquitous antioxidant in green tea, was able to prevent the UV-induced skin damage [34].

Similarly, *Ginkgo biloba* extracts are found to be rich in flavonoids such as rutin, quercetin, and kaempferol, thus having antioxidant properties and subsequently displayed photoprotective potency [21]. The abundance of flavonoids as well as biflavones and terpene trilactones is also responsible for anti-inflammatory and vasodilatory effects in *G. biloba* [21]. The *in vivo* tests conducted on hairless mice revealed that the application of formulations containing a glycolic extract of *G. biloba* could protect against UV-induced skin damage (erythema, sunburn cell formation). Further, the results indicated that those photoprotective effects might not solely due to the UV-absorbance by UV filters, but due to the biological effects induced upon the application of the formulations comprising *G. biloba* extracts [35].

Resveratrol is a polyphenol abundant in grapes and red wine, possessing numerous biological activities such as antioxidant, anti-inflammatory, anti-proliferative, and cardioprotective effects. Zhou et al. [36] demonstrated that pretreatment of resveratrol could increase survival of UV-B-treated human keratinocyte cell line HaCaT cells while reducing the generation of ROS. Further, this pre-treatment had a direct impact on the caspase pathway, hence cell survival [36].

Curcumin derived from rhizomes of *Curcuma longa* (turmeric) has inherited a plethora of biological activities like anti-inflammatory, antioxidant, anticarcinogenic as well as anti-infective potencies. Topical application of curcumin is found to be an effective strategy to prevent and/or treat UV-induced acute inflammation and photoaging [37]. It was also observed that the topical application of curcumin prior to UV-B irradiation in hairless mice skin could reduce inflammation-related parameters like infiltration of inflammatory cells, collagen accrementition derangement as well as lipid peroxidation. It has also induced nuclear factor Nrf2 which functions as a regulator in the defense against oxidative stress. Moreover the treatment of HaCaT cells with curcumin prior to acute UV-B irradiation has resulted in reduced leakage of lactate dehydrogenase (LDH) indicating minimal cell damage. Further, curcumin is capable of effectively scavenging ROS generated due to UV-B exposure [37].

A lyophilized extract obtained from a methanolic extraction of the flowering buds of *Capparis spinosa* (Capers) was found to be rich in phenolic acids (caffeic acid, ferulic acid, p-coumaric acid) and flavonoids (kaempferol and quercetin derivatives). These polyphenolic compounds are known to have prominent antioxidant effects. Further, the application of a gel formulation constituting the aforementioned extract on healthy human volunteers demonstrated an inhibitory effect against UV-B-induced erythema [38].

Garcinia brasiliensis is a plant that has been found throughout Brazil and also native to the Amazon forest. The fruit of this plant is used as a folk medicine to treat peptic ulcers, urinary, and tumor diseases. The plants of the *Garcinia* species have received considerable attention during the past few decades as photoprotective agents due to the presence of polyphenols such as bioflavonoids. Furthermore, the extracts of the pericarp, epicarp, and seeds of *Garcinia* fruit proved the antioxidant and anti-inflammatory activities. Based on that, Figueiredo et al. [39] investigated the *in vitro* and *in vivo* photoprotective activity in the ethanolic extract of the epicarp of the fruits. Their results revealed that the extract had a higher potential to be used as a sunscreen in cosmetic formulations instead of synthetic UV filters. Especially, the *in vivo* test results demonstrated the photoprotective effect of the extract by decreasing UV-B-induced damages such as the secretion of the pro-inflammatory cytokines [39].

Costa et al. [40] observed the *in vitro* photoprotective activity of an ethanolic extract of the leaves of *Marcetia taxifolia*, an endemic plant grown in Northeastern Brazil, and the potential use of that extract in sunscreen formulations. The results showed the photoprotective activity of the extract against UV-A and UV-B radiations, while obtaining SPF value of 15.52 at a concentration of 250 µg/µL. The sunscreen formulations prepared from the same extract displayed SPF values ≥ 6 indicating the suitability of the extract to use as an active ingredient for sunscreens [40]. In another study, sunscreen delivery systems comprising rutin, *Passiflora incarnata*, and *Plantago lanceolata* along with organic/inorganic UV filters has significantly enhanced the protection against UV-A radiation [41].

Some vegetables as well had exhibited the sun protective capability. According to Mazumder et al. [42], the *in vitro* assay of hydroalcoholic extracts of beet root, green pea, drumstick, and sweet potato had recorded high SPF values [42]. Patil et al. [43] had witnessed a sunscreen comprising *Pongamia pinnata* and *Punica granatum* extracts in a 3:2 ratio showing effective sunscreen potential [43]. A combination of dried rosemary (*R. officinalis*) leaf extract and grapefruits (*Citrus paradisi*) extract appeared as a highly effective photoprotective agent in skin cell model and in a human clinical study [44]. A clinical study performed by Egoumenides et al. [45] indicated that the application and/or supplementation of melon concentrate could increase minimal erythema dose (MED – the amount of UV radiation that produces minimal erythema of an individual's skin within a few hours following exposure) and also the endogenous antioxidant enzymes while decreasing the sun burn cells [45]. Water dropwort (*Oenanthe javanica*), a herb consumed as a vegetable and a medicinal plant, was found to be effective against UV-B-induced collagen disruption and inflammation in mouse model. It decreased the expression of matrix metalloproteinases (MMP), tumor necrosis factor (TNF)-α, and cyclooxygenase (COX)-2 while increasing the productions of collagen types I and III and suggested the potential usefulness in the treatment of photodamaged skin [46].

Similarly, the fruit extract of *Sechium edule* has displayed photoprotection potential against UV-A in human keratinocytes. In the presence of the extract, the keratinocytes have been protected against UV-A-induced cytotoxicity and concurrently the intracellular amounts of ROS were drastically reduced [47]. Cefali et al. [48] reported a flavonoid-enriched phytocosmetic sunscreen formulation, an (O/W) emulsion with a blend of plant extracts obtained from the leaves of *Ginkgo biloba*, *Dimorphandra mollis*, *Ruta graveolens*, and *Vitis vinifera*. The prepared sunscreen formulation was proved to be effective against both UV-A and UV-B radiation and found to be non-irritant to skin [48].

The extract of *Polypodium leucotomos* was also found to be effective in blocking the deleterious effects of UV irradiation. This was observed in both *in vivo* and *in vitro* models. The extract could inhibit free radical generation, preventing photodecomposition of endogenous photoprotective molecules and DNA and ultimately to prevent the UV-induced cell death [49].

Euterpe oleracea is a native palm of the Amazon region rich with anthocyanins. Daher et al. [20] prepared oil-water sunscreens emulsions using glycolic extract of the fruit and the resulting emulsion displayed significant UV-A and UV-B protectant effects [20]. In another study sunscreen formulations were prepared with *Helichrysum arenarium*, *Sambucus nigra*, and *Crataegus monogyna* extracts in which phenolic compounds, specially flavonoids, are abundant. The *in vitro* photoprotective potential and photostability were determined. All the emulsions prepared in the study displayed good UV protection and photostability while the activity was more prominent in the emulsion comprised extract combination [50].

The leaves of *Buddleja scordioides* displayed *in vivo* photoprotective properties. This plant is widely spread throughout Mexico, and in folklore medicine it is used to treat tumors, abscesses, sores, and burns. A methanolic extract of the leaves of the plant could significantly lower the UV-B-induced erythema of female SKH-1 hairless mice. The assessment was done after the topical application of methanolic extract on the skin of the mice. The *in vivo* test results were supported by the *in vitro* test results. During the *in vitro* tests, the extract reported high absorbance within the UV-B range, and this was attributed to the presence of polyphenols, mainly verbascoside and linarin [51]. In addition there are many recent reports on plants with *in vitro* photoprotective potential. This includes plants like *Neoglaziovia variegata* [52], *Moringa oleifera* [53], *Aloe vera* [54], and *Juglans regia* [55].

The Sri Lankan flora has been rarely examined for photoprotective molecules. Napagoda et al. [5] identified six Sri Lankan medicinal plants with high photoprotective activity. The methanol-water extracts of the leaves of *Atalantia ceylanica*, *Hibiscus furcatus*, *Olax zeylanica*, and *Ophiorrhiza mungos* have displayed SPF values ≥ 25, suggesting an advanced UV-B filtering activity in the aforementioned extracts. Moreover, the extracts prepared from seeds of *Mollugo cerviana* and whole plants of *Leucas zeylanica* recorded SPF values of 29.5 and 39.8 respectively. Interestingly, the UV absorption profiles of the aqueous-methanolic extracts of *L. zeylanica* (Figure 8.2 (A)) and *O. mungos* (Figure 8.2 (B)) displayed the properties that required for an ideal sunscreen owing to the high absorbance values within both UV-A range and UV-B range [5].

(A)   **(B)**

Figure 8.2: (A) *Leucas zeylanica* (B) *Ophiorrhiza mungos*.

8.4 Increasing the efficacy of herbal sunscreens

Cosmetic industry is using nanotechnology to enhance the bioavailability, permeability, and stability of sunscreen formulations. The efficiency of herbal photoprotective extracts had enhanced when formulated as nanopreparations [56].

The addition of TiO_2 and ZnO nanoparticles into herbal sunscreens is beneficial for poorly soluble, poorly absorbed, and labile herbal extracts and phytochemicals in them. Besides, the sunscreens prepared with nanoparticles are non-greasy, of low viscosity, and high bioavailability. Furthermore, the presence of these nanoparticles in sunscreens can improve the stability of chemically unstable active ingredients, control the release of active ingredients, and improve skin hydration and protection by film formation on the skin [57]. Although herbal sunscreen formulations produced through the combination of nanoparticles and herbal extracts appears to be very innovative, there are only a few reports on success stories. In an early attempt where quercetin and rutin were used in association with TiO_2, synergistic effect was observed with a significant increase in the SPF values [24]. In another study, zinc oxide nanoparticles were incorporated into lyophilized methanolic extract of the top flowerings of *Teucrium polium* which had significantly enhanced the SPF value while decreasing the photodecomposition of the compounds in the sunscreen formulation [58]. Recently, Battistin et al. [59] reported the functionalization of TiO_2 nanoparticles with the polyphenolic antioxidant molecule, oxisol. The results revealed that the antioxidant potential of a sunscreen formulation could be enhanced due to the high surface area offered by the TiO_2 nanoparticles to the organic molecules [59].

Incorporation of herbal sunscreens into solid lipid nanoparticles (SLNs) gives additional protection against UV radiation because highly crystalline SLNs can reflect and scatter the UV radiation and act as physical UV blockers [60]. SLNs can be prepared in many ways such as hot homogenization method, cold homogenization method, and spray drying method. These SLNs increase the stability of the UV blockers in herbal sunscreens and permit the extended release of UV blockers into the skin giving an improved photoprotection capability to the herbal sunscreen [61]. For example, SLNs-based sunscreen formulation comprising the natural antioxidant molecule quercetin has displayed better localization of the active ingredient within the skin in comparison to the formulation with particles in the micrometer range. This kind of accumulation of quercetin in the skin is highly beneficial in delaying UV-mediated cellular damage [62]. In another study, SLNs and nanostructured lipid carriers (NLC) containing resveratrol displayed antioxidant properties, however, resveratrol-loaded NLC appeared to be superior to its counterpart as it is capable of penetrating deeper into the skin [63]. The *Aloe vera*–loaded SLN sunscreen formulation was also found to be effective and displayed good SPF value that was on par with the sunscreens available in the market [64]. Similarly, sunscreen formulations containing silymarin SLNs [65], polymeric nanoparticles of naringenin [66], and polymeric nanoparticles of morin [67] have also exhibited excellent photoprotective activities.

Polyphenols incorporated in herbal sunscreens are very unstable without any encapsulation. Therefore nanoencapsulation is often used to increase the long-term stability and bioavailability [68]. Further, chitosan nanoparticles were used as a nanocarrier to formulate sunscreen emulsions with saffron (*Crocus sativus*) and annatto (*Bixa orellana*) [69] while a photoprotective antioxidant nanoemulsion containing chitosan and pomegranate extract was found to be highly effective and photostable [70]. In another study, photoprotective capacity in the strawberry extracts were potentiated upon the development of strawberry-based formulations with nanoparticles of Coenzyme Q10 [71]. Further Bucci et al. [72] developed "nanoberries," the ultradeformable liposomes enriched with blueberry (*Vaccinium myrtillus*) that are inherited with differential elastic properties, capable of penetrating stratum corneum and exerting low toxicity. Interestingly, in the presence of nanoberries, the viability of HaCaT cells was maintained even upon the exposition to UV radiation [72].

Based on the foregoing discussion, it is fair to assume that nano-based approaches are inevitable to develop highly effective herbal sunscreen formulations.

8.5 Conclusion

Exposure to UV radiation in sunlight poses several health risks and invites preventive measures to be taken to avoid those consequences. Topical application of sunscreens would be the best option as a preventive measure. Over the years, sunscreens of herbal origin are gaining popularity, and the research findings show that the efficacy of these herbal sunscreen products can be enhanced using nanotechnology.

References

[1] Pelizzo M, Zattra E, Nicolosi P, Peserico A, Garoli D, Alaibac M. *In vitro* evaluation of sunscreens: An update for the clinicians. Int Sch Res Notices 2012, 2012, 352135. doi:10.5402/2012/352135

[2] Jansen R, Wang SQ, Burnett M, Osterwalder U, Lim HW. Photoprotection: Part I. Photoprotection by naturally occurring, physical, and systemic agents. J Am Acad Dermatol 2013, 69(6), 853. e1–866.

[3] Clydesdale GJ, Dandie GW, Muller HK. Ultraviolet light induced injury: Immunological and inflammatory effects. Immunol Cell Biol 2001, 79(6), 547–568.

[4] Curnow A, Owen SJ. An evaluation of root phytochemicals derived from *Althea officinalis* (marshmallow) and *Astragalus membranaceus* as potential natural components of UV protecting dermatological formulations. Oxid Med Cell Longev 2016, 2016, 7053897. 10.1155/2016/7053897.

[5] Napagoda MT, Malkanthi BM, Abayawardana SA, Qader MM, Jayasinghe L. Photoprotective potential in some medicinal plants used to treat skin diseases in Sri Lanka. BMC Complement Altern Med 2016, 16(1), 479. 10.1186/s12906-016-1455-8

[6] Scharffetter-Kochanek K, Wlaschek M, Brenneisen P, Schauen M, Blaudschun R, Wenk J, UV-induced reactive oxygen species in photocarcinogenesis and photoaging. Biol Chem 1997, 378(11), 1247–1257.

[7] Liu Z, Ren Z, Zhang J, Chuang CC, Kandaswamy E, Zhou T, Zuo L. Role of ROS and nutritional antioxidants in human diseases. Front Physiol 2018, 9, 477. 10.3389/fphys.2018.00477.

[8] Prasad A, Pospíšil P. Ultraweak photon emission induced by visible light and UltraViolet A radiation via photoactivated skin chromophores: *In vivo* charge coupled device imaging. J Biomed Opt 2012, 17(8), 085004. 10.1117/1.JBO.17.8.085004.

[9] Rinnerthaler M, Bischof J, Streubel MK, Trost A, Richter K. Oxidative stress in aging human skin. Biomolecules. 2015, 5(2), 545–589. 10.3390/biom5020545.

[10] Chen L, Hu JY, Wang SQ. The role of antioxidants in photoprotection: A critical review. J Am Acad Dermatol 2012, 67(5), 1013–1024.

[11] Fioletov V, Kerr JB, The FA. UV index: Definition, distribution and factors affecting it. Can J Public Health 2010, 101(4), I5–19.

[12] WHO 2002. Global Solar UV Index: A Practical Guide. [Online] Available at: https://www.who.int/uv/publications/en/UVIGuide.pdf [Accessed 23 May 2021]

[13] https://www.epa.gov/sunsafety/uv-index-scale-0 [Accessed 4 Feb 2019]

[14] https://www.medicinenet.com [Accessed 4 Feb 2020]

[15] Gasparro FP, Mitchnick M, Nash JF. A review of sunscreen safety and efficacy. Photochem Photobiol 1998, 68(3), 243–256.

[16] Serpone N, Dondi D, Albini A. Inorganic and organic UV filters: Their role and efficacy in sunscreens and suncare products. Inorganica Chim Acta 2007, 360(3), 794–802.

[17] Burnett ME, Wang SQ. Current sunscreen controversies: A critical review. Photodermatol Photoimmunol Photomed 2011, 27(2), 58–67.

[18] Manaia E, Kaminski R, Corrêa M, Inorganic CL. UV filters. Braz J Pharm Sci 2013, 49(2), 201–209.

[19] Nichols JA, Katiyar SK. Skin photoprotection by natural polyphenols: Anti-inflammatory, antioxidant and DNA repair mechanisms. Arch Dermatol Res 2010, 302, 71–83.

[20] Daher CC, Fontes IS, Rodrigues RO, GAdeB D, DdosS S, Cfs A, Gomes APB, Ferrari M. Development of O/W emulsions containing *Euterpe oleracea* extract and evaluation of photoprotective efficacy. Braz J Pharm Sci 2014, 50, 639–652.

[21] Cefali LC, Ataide JA, Moriel P, Foglio MA, Mazzola PG. Plant-based active photoprotectants for sunscreens. Int J Cosmet Sci 2016, 38(4), 346–353.

[22] Klein FRS, Reis A, Kleinowski AM, Telles RT, Amarante LD, Ja P, Ejb B. UV-B radiation as an elicitor of secondary metabolite production in plants of the genus *Alternanthera*. Acta Bot Bras 2018, 32(4), 615–623.

[23] Nocchi N, Duarte HM, Pereira RC, Konno TUP, Soares AR. Effects of UV-B radiation on secondary metabolite production, antioxidant activity, photosynthesis and herbivory interactions in *Nymphoides humboldtiana* (Menyanthaceae). J Photochem Photobiol B Biol 2020, 212, 112021. 10.1016/j.jphotobiol.2020.112021.

[24] Choquenet B, Couteau C, Paparis E, Coiffard LJM Quercetin and rutin as potential sunscreen agents: Determination of efficacy by an *in vitro* method. J Nat Prod 2008, 71(6), 1117–1118.

[25] Wei H, Saladi R, Lu Y, Wang Y, Palep SR, Moore J, Phelps R, Shyong E, Lebwohl MG. Isoflavone genistein: Photoprotection and clinical implications in dermatology. J Nutr 2003, 133, 3811S–3819S.

[26] Douglas CJ. Phenylpropanoid metabolism and lignin biosynthesis: From weeds to trees. Trends Plant Sci 1996, 1, 171–178.

[27] Korkina L, Kostyuk V, Potapovich A, Mayer W, Talib N, De Luca C. Secondary plant metabolites for sun protective cosmetics: From pre-selection to product formulation. Cosmetics 2018, 5, 32. 10.3390/cosmetics5020032

[28] Afaq F. Natural agents: Cellular and molecular mechanisms of photoprotection. Arch Biochem Biophys 2011, 508(2), 144–151.

[29] Sánchez-Campillo M, Gabaldon JA, Castillo J, Benavente-García O, Del Baño MJ, Alcaraz M, Vicente V, Alvarez N, Lozano JA. Rosmarinic acid, a photo-protective agent against UV and other ionizing radiations. Food Chem Toxicol 2009, 47(2), 386–392.

[30] Katiyar SK, Perez A, Mukhtar H. Green tea polyphenol treatment to human skin prevents formation of UltraViolet light B-induced pyrimidine dimers in DNA. Clin Cancer Res 2000, 6, 3864–3869.

[31] Afaq F, Mukhtar H. Botanical antioxidants in the prevention of photocarcinogenesis and photoaging. Exp Dermatol 2006, 15(9), 678–684.

[32] Kaur CD, Saraf S. Photochemoprotective activity of alcoholic extract of *Camellia sinensis*. Int J Pharmacol 2011, 7(3), 400–404.

[33] Heinrich U, Moore CE, De Spirt S, Tronnier H, Stahl W. Green tea polyphenols provide photoprotection, increase microcirculation, and modulate skin properties of women. J Nutr 2011, 141(6), 1202–1208.

[34] Jeon HY, Kim JK, Kim WG, Lee SJ. Effects of oral epigallocatechin gallate supplementation on the minimal erythema dose and UV-induced skin damage. Skin Pharmacol Physiol 2009, 22(3), 137–141.

[35] Dal Belo SE, Gaspar LR, Campos PMBGM. Photoprotective effects of topical formulations containing a combination of *Ginkgo biloba* and green tea extracts. Phyt Res 2011, 25, 1854–1860.

[36] Zhou F, Huang X, Pan Y, Cao D, Liu C, Liu Y, Chen A. Resveratrol protects HaCaT cells from ultraviolet B-induced photoaging via upregulation of HSP27 and modulation of mitochondrial caspase-dependent apoptotic pathway. Biochem Biophys Res Commun 2018, 499(3), 662–668.

[37] Li H, Gao A, Jiang N, Liu Q, Liang B, Li R, Zhang E, Li Z, Zhu H. Protective effect of curcumin against acute Ultraviolet B irradiation-induced photo-damage. Photochem Photobiol 2016, 92(6), 808–815.

[38] Bonina F, Puglia C, Ventura D, Aquino R, Tortora S, Sacchi A, Saija A, Tomaino A, Pellegrino ML, de Caprariis P. *In vitro* antioxidant and *in vivo* photoprotective effects of a lyophilized extract of *Capparis spinosa* L buds. J Cosmet Sci 2002, 53(6), 321–335.

[39] Figueiredo SA, Vilela FM, Da Silva CA, Cunha TM, Dos Santos MH, Fonseca MJ. *In vitro* and *in vivo* photoprotective/photochemopreventive potential of *Garcinia brasiliensis* epicarp extract. J Photochem Photobiol B 2014, 131, 65–73.

[40] Costa CCS, Detoni CB, Branco CRC, Botura MB, Branco A. *In vitro* photoprotective effects of *Marcetia taxifolia* ethanolic extract and its potential for sunscreen formulations. Rev Bras Farmacogn 2015, 25, 413–418.

[41] Velasco MV, Sarruf FD, Salgado-Santos IM, Haroutiounian-Filho CA, Kaneko TM, Baby AR. Broad spectrum bioactive sunscreens. Int J Pharm 2008, 363(1–2),50–57.

[42] Mazumder MU, Das K, Choudhury AD, Khazeo P. Determination of sun protection factor (SPF) number of some hydroalcoholic vegetable extracts. PharmaTutor 2018, 6(12), 41–45.

[43] Patil S, Fegade B, Zamindar U, Bhaskar VH. Determination of sun protection effect of herbal sunscreen cream. WJPPS 2015, 4(8),1554–1565.

[44] Nobile V, Michelotti A, Cestone E, Caturla N, Castillo J, Benavente-García O, Pérez-Sánchez A, Micol V. Skin photoprotective and antiageing effects of a combination of rosemary (*Rosmarinus officinalis*) and grapefruit (*Citrus paradisi*) polyphenols. Food Nutr Res 2016, 60, 31871. 10.3402/fnr.v60.31871.

[45] Egoumenides L, Gauthier A, Barial S, Saby M, Orechenkoff C, Simoneau G, Carillon J. A specific melon concentrate exhibits photoprotective effects from antioxidant activity in healthy adults. Nutrients 2018, 10(4), 437. 10.3390/nu10040437.

[46] Her Y, Shin BN, Lee YL, Park JH, Kim DW, Kim KS, Kim H, Song M, Kim JD, Won MH, Ahn JH. *Oenanthe Javanica* extract protects mouse skin from UVB radiation via attenuating collagen disruption and inflammation. Int J Mol Sci 2019, 20(6), 1435. 10.3390/ijms20061435.

[47] Metral E, Rachidi W, Damour O, Demarne F, Bechetoille N. Long-term genoprotection effect of *Sechium edule* fruit extract against UVA irradiation in keratinocytes. Photochem Photobiol 2018, 94, 343–350.

[48] Cefali LC, Ataide JA, Fernandes AR, Sousa IMO, Fcds G, Eberlin S, Dávila JL, Jozala AF, Chaud MV, Sanchez-Lopez E, Marto J, d'Ávila MA, Ribeiro HM, Foglio MA, Souto EB, Mazzola PG. Flavonoid-enriched plant-extract-loaded emulsion: A novel phytocosmetic sunscreen formulation with antioxidant properties. Antioxidants (Basel) 2019, 8(10), 443. 10.3390/antiox8100443.

[49] Gonzalez S, Alonso-Lebrero JL, Del Rio R, Jaen P. *Polypodium leucotomos* extract: A nutraceutical with photoprotective properties. Drugs Today (Barc). 2007, 43(7), 475–485.

[50] Jarzycka A, Lewińska A, Gancarz R, Wilk KA. Assessment of extracts of *Helichrysum arenarium*, *Crataegus monogyna*, *Sambucus nigra* in photoprotective UVA and UVB; photostability in cosmetic emulsions. J Photochem Photobiol B 2013, 128, 50–57. 10.1016/j.jphotobiol.2013.07.029.

[51] Avila Acevedo JG, Espinosa González AM, De Maria Y Campos DM, Benitez Flores Jdel C, Hernández Delgado T, Flores Maya S, Campos Contreras J, Muñoz López JL, García Bores AM. Photoprotection of *Buddleja cordata* extract against UVB-induced skin damage in SKH-1 hairless mice. BMC Complement Altern Med 2014, 14, 281. 10.1186/1472-6882-14-281.

[52] Oliveira Junior R, CdeS A, Gr S, Al G, APde O, Lima-Saraiva SRG, Acs M, JSRdos S, JRGdaS A. *In vitro* antioxidant and photoprotective activities of dried extracts from *Neoglaziovia variegata* (Bromeliaceae). J App Pharm Sci 2013, 3 (1), 122–127.

[53] Baldisserotto A, Buso P, Radice M, Dissette V, Lampronti I, Gambari R, Manfredini S, Vertuani S. *Moringa oleifera* leaf extracts as multifunctional ingredients for "natural and organic" sunscreens and photoprotective preparations. Molecules. 2018, 23(3),664. 10.3390/molecules23030664.

[54] Rodrigues D, Viotto AC, Checchia R, Gomide A, Severino D, Itri R, Baptista MS, Martins WK. Mechanism of *Aloe vera* extract protection against UVA: Shelter of lysosomal membrane avoids photodamage. Photochem Photobiol Sci 2016, 15(3), 334–350.

[55] Muzaffer U, Paul V, Prasad NR, Karthikeyan R, Agilan B. Protective effect of *Juglans regia* L. against ultraviolet B radiation induced inflammatory responses in human epidermal keratinocytes. Phytomedicine 2018, 42, 100–111.

[56] Sahu AN. Nanotechnology in herbal medicines and cosmetics. Int J Res Ayurveda Pharm 2013, 4(3), 472–474

[57] Chanchal D, Swarnlata S. Novel approaches in herbal cosmetics. J Cosmet Dermatol 2008, 7(2), 89–95.

[58] Sharififar F, Ansari M, Kazemipour M, Mahdavi H, Sarhadinejad Z. *Teucrium polium* L. extract adsorbed on zinc oxide nanoparticles as a fortified sunscreen. Int J Pharm Investig 2013, 3(4), 188–193.

[59] Battistin M, Dissette V, Bonetto A, Durini E, Manfredini S, Marcomini A, Casagrande E, Brunetta A, Ziosi P, Molesini S, Gavioli R, Nicoli F, Vertuani S, Baldisserotto A. A new approach to UV protection by direct surface functionalization of TiO_2 with the antioxidant polyphenol dihydroxyphenyl benzimidazole carboxylic acid. Nanomaterials 2020, 10(2), 231. 10.3390/nano10020231

[60] Jain SK, Jain NK. Multiparticulate carriers for sun-screening agents. Int J Cosmet Sci 2010, 32(2), 89–98.

[61] Jose J, Netto G. Role of solid lipid nanoparticles as photoprotective agents in cosmetics. J Cosmet Dermatol. 2019, 18(1), 315–321.

[62] Bose S, Du Y, Takhistov P, Michniak-Kohn B. Formulation optimization and topical delivery of quercetin from solid lipid based nanosystems. Int J Pharm. 2013, 441(1–2), 56–66.

[63] Gokce EH, Korkmaz E, Dellera E, Sandri G, Bonferoni MC, Ozer O. Resveratrol-loaded solid lipid nanoparticles versus nanostructured lipid carriers: Evaluation of antioxidant potential for dermal applications. Int J Nanomedicine 2012, 7, 1841–1850.

[64] Rodrigues LR, Jose J. Exploring the photo protective potential of solid lipid nanoparticle-based sunscreen cream containing *Aloe vera*. Environ Sci Pollut Res Int 2020, 27(17), 20876–20888.

[65] Netto G, Jose J. Development, characterization, and evaluation of sunscreen cream containing solid lipid nanoparticles of silymarin. J Cosmet Dermatol 2018, 17(6), 1073–1083.

[66] Joshi H, Hegde AR, Shetty PK, Gollavilli H, Managuli RS, Kalthur G, Mutalik S. Sunscreen creams containing naringenin nanoparticles: Formulation development and *in vitro* and *in vivo* evaluations. Photodermatol Photoimmunol Photomed 2018, 34(1), 69–81.

[67] Shetty PK, Venuvanka V, Jagani HV, Chethan GH, Ligade VS, Musmade PB, Nayak UY, Reddy MS, Kalthur G, Udupa N, Rao CM, Mutalik S. Development and evaluation of sunscreen

creams containing morin-encapsulated nanoparticles for enhanced UV radiation protection and antioxidant activity. Int J Nanomedicine 2015, 10, 6477–6491.

[68] Munin A, Edwards-Lévy F. Encapsulation of natural polyphenolic compounds; a review. Pharmaceutics. 2011, 3(4), 793–829.

[69] Ntohogian S, Gavriliadou V, Christodoulou E, Nanaki S, Lykidou S, Naidis P, Mischopoulou L, Barmpalexis P, Nikolaidis N, Bikiaris DN. Chitosan nanoparticles with encapsulated natural and uf-purified annatto and saffron for the preparation of UV protective cosmetic emulsions. Molecules 2018, 23(9), 2107. 10.3390/molecules23092107.

[70] Cerqueira-Coutinho C, Santos-Oliveira R, Dos Santos E, Mansur CR. Development of a photoprotective andantioxidant nanoemulsion containing chitosanas an agent for improving skin retention. Eng Life Sci 2015, 15, 593–604.

[71] Gasparrini M, Forbes-Hernandez TY, Afrin S, Alvarez-Suarez JM, Gonzàlez-Paramàs AM, Santos-Buelga C, Bompadre S, Quiles JL, Mezzetti B, Giampieri F. A pilot study of the photoprotective effects of strawberry-based cosmetic formulations on human dermal fibroblasts. Int J Mol Sci 2015, 16(8), 17870–17884.

[72] Bucci P, Prieto MJ, Milla L, Calienni MN, Martinez L, Rivarola V, Alonso S, Montanari J. Skin penetration and UV-damage prevention by nanoberries. J Cosmet Dermatol. 2018, 17(5), 889–899.

Mayuri Napagoda

9 Poisonous plants and their toxic metabolites

9.1 Introduction

Since the beginning of the evolution of land plants in the Devonian period, plants faced many challenges from their surroundings, for example, unfavorable environmental conditions, herbivore attacks and infections caused by bacteria, fungi, and viruses [1–3]. As survival mechanisms, plants produce a diverse array of organic compounds which are called "secondary metabolites." These include alkaloids, amines, cyanogenic glycosides, glucosinolates, non-protein amino acids, organic acids, terpenoids, phenols, quinones, and several other classes of secondary metabolites [3]. Rather than synthesizing a single class of secondary metabolites for defense purposes, plants usually produce mixtures of secondary metabolites of different structural classes, and this could result in additive or synergistic effects. Some of these metabolites exert toxic effects on herbivore pests, livestock animals, and even on humans [4]. These toxic metabolites could attack different molecular targets in the victim, for example, ion channels, neurotransmitter receptors, neurotransmitter inactivating enzymes, cellular respiration in mitochondria, protein biosynthesis in ribosomes, DNA, and RNA [3, 4]. Depending on the growth stage and part of the plant, the quantity consumed, the species, and the susceptibility of the victim, toxic effects could vary [5].

9.2 The use of poisonous plants: some historical reports

Since ancient times poisonous plants have been utilized for various purposes, not only as food and medicines but also as agents for crime, punishment, suicide, and bioterrorism. Some of these plants have been used as arrow and dart poisons as well as fish poisons by various tribal communities, and even for recreational and spiritual purposes [6].

The prevalence of toxic substances in plants might have severely restricted the utility of vegetables and fruits as food for primitive man. However, with the use of fire for cooking, it is believed that many of these substances get removed, thus increasing the palatability of plant materials to humans [7]. Some examples of edible

Mayuri Napagoda, Faculty of Medicine, University of Ruhuna, Galle, 80 000, Sri Lanka,
e-mail: mayurinapagoda@yahoo.com

https://doi.org/10.1515/9783110595949-009

poisonous plants include *Manihot esculenta* (cassava/ manioc) and cycads like *Cycas circinalis* which are consumed after proper processing [6].

Some toxic compounds present in plants have displayed a wide range of pharmaceutical properties depending on the dose administered, thus developed into lifesaving therapeutic agents. For example, d-tubocurarine obtained from *Chondrodendron tomentosum* is used as a muscle relaxant in anesthesia [8] while the glycosides isolated from *Digitalis* spp. are used in the management of congestive heart failure and cardiac arrhythmias [9]. Moreover, camptothecin isolated from *Camptotheca acuminata* as well as vincristine and vinblastine from *Catharanthus roseus* have been utilized as anti-cancer drugs [10]. Another example is colchicine extracted from *Colchicum autumnale*. It is used in the treatment of conditions like gout and familial Mediterranean fever [11].

There is historical evidence on the use of poisonous plant species in capital punishment. A well-known example is the execution of Socrates, who was forced to drink a cup of poison hemlock (*Conium maculatum*) in 399 BCE [12]. In many African communities, "trial by ordeal" was a judicial practice used to determine whether the accused was guilty or innocent by subjecting them to an unpleasant experience. Various plant species were used as ordeal poisons and *Physostigma venenosum*, *Tanghinia venenifera* (a synonym of *Cerbera manghas*), *Menabea venenata*, and *Strychnos icaja* are to name a few [6, 13].

Ricin is a toxic protein (toxalbumin) produced in the seeds of the castor oil plant, *Ricinus communis*. Despite its extensive utility in traditional medicine, it is also used for criminal purposes. One of the notable examples is the assassination of a communist dissident Georgi Markov in 1978 [14].

In the traditional societies, extracts prepared from poisonous plants were applied on arrows, darts, or javelins and those weapons were used in hunting as well as to protect from wild animals, or for martial purposes. These include plants like *Calotropis procera*, *Clathrotropis glaucophylla*, *Strychnos guianensis*, *Antiaris toxicaria*, *Maquira* spp., and *Naucleopsis* spp. [15]. Plant toxins were also used in fishing and even today it is practiced in some remote areas in the African continent. *Prosopis africana*, *Quassia africana*, *Euphorbia* spp., *Strychnos* spp. are some examples and the saponins, rotenoids, and diterpene esters are identified as compounds mainly responsible for piscicidal activity [16].

A number of plant secondary metabolites are used as insect antifeedants or repellents, fungicides, bactericides, molluscicides, nematicides, and rodenticides [6]. Plants like *Derris*, *Lonchocarpus* and *Tephrosia* that have been used as fishing poisons produce rotenone which also display insecticidal properties [17]. Strychnine isolated from *Strychnos* spp. is another example of a botanical pesticide [6].

Since ancient times, many plant species have been used in religious rituals and as hallucinogens, stimulants, or sedatives. For example, *Papaver somniferum* (opium poppy) and several other species of Papaveraceae have been widely used for sedation and pain relief while *Datura stramonium, Atropa belladonna, Cannabis* spp., and *Mandragora* spp. display hallucinogenic, sedative, and aphrodisiac properties [6].

Some plant species contain carcinogenic and/or teratogenic substances. Chronic ingestion of bracken fern (*Pteridium aquilinum*) could exert toxic effects in cattle leading to bovine enzootic hematuria (BEH) and carcinoma of the esophagus [18]. Aristolochic acid present in *Aristolochia* spp. has been identified as a genotoxic mutagen and is believed to be associated with the development of nephropathy and urothelial cancer [19].

9.3 Some common poisonous plants in the Indian subcontinent

A large number of poisonous plants grow wild in the Indian subcontinent and are identified as an important cause of mortality among adults [20, 21]. Besides, plant poisoning in children is considered as one of the common presentations to emergency departments [21]. Numerous reports revealed that small children have unintentionally eaten plants with attractive fruits or flowers. Self-poisoning with plant seeds/fruits or misidentification as edible plants is responsible for most of the fatalities among adults [22, 23].

Given below are some poisonous plant species widely available in the Indian subcontinent.

9.3.1 *Datura stramonium* (thorn apple/jimson weed)

Datura species, particularly *D. stramonium*, are rich in belladonna alkaloids like atropine, scopolamine, and hyoscyamine. These compounds may exert both local and systemic anticholinergic toxicity. *D. stramonium* is reputed for its hallucinogenic properties, and the poisoning could occur due to the overdosing of the herbal medication as well as accidental food contamination [24, 25]. In addition, the voluntary ingestion of *D. stramonium* for its hallucinogenic and euphoric effects was the common cause for *Datura* poisoning among teenagers [26].

Over 28 belladonna alkaloids have been identified in different *Datura* species, and the ratio of these alkaloids varies from species to species and also among the specimens of the same species. As a result, different profiles of toxicity are observed in patients. Although these alkaloids are distributed in all parts of the plant, the highest content is found in the petioles (flowers) and the least amount in the roots. The toxicity results in a blockade of peripheral muscarinic receptors causing mydriasis, dry mouth, tachycardia, fever, and erythema [24].

9.3.2 *Tabernaemontana dichotoma* (synonym: *Pagiyantha dicotoma*)

Tabernaemontana dichotoma is locally called Eve's apple. Although it is used in traditional medicine to treat wounds, snake bites, and ulcers, it is listed among the most poisonous plants in Sri Lanka.

The fruits of the plant are highly poisonous and the seeds are found to be narcotic, producing delirium. The leaves and the stem bark display purgative properties. The plant is rich in alkaloids; for example, coronaridine, tabersonine, heyneanine, voacristine hydroxyindolenine, perivine, 19-epivoacristine, 12-methoxy-voaphylline, and vobasine were isolated from different parts of this plant [27].

9.3.3 *Strychnos nux vomica* (nux vomica)

The seeds of *Strychnos nux vomica* are used in traditional medicine; however, due to the presence of toxic alkaloids like strychnine and brucine, the plant is categorized as a poisonous plant. These toxic compounds could be found in the entire plant but distributed in high concentrations in seeds and bark [28, 29].

Strychnine and brucine are neurotoxins and function as competitive antagonists of the glycine receptors on the postsynaptic membrane in the spinal cord, brain stem, etc. The ingestion of lethal doses of strychnine may result in respiratory or spinal paralysis, cardiac arrest, and ultimately death. Although brucine alone is not as toxic as strychnine, still the lethal dose of this alkaloid may cause rhabdomyolysis and acute renal failure [30]. In some countries, strychnine is used to control wild/stray dogs and foxes. The poisoning may occur due to overdosing of native herbal formulations and as adulterants in illicit drugs or as attempts of deliberate self-harm by consumption of plant parts or rodenticides [29].

9.3.4 *Cerbera manghas* (sea mango)

Cerbera species, particularly *Cerbera manghas*, are distributed throughout the tropical coasts. The kernel of this plant is identified as the most toxic part mainly due to the presence of several cardiac glycosides [31]. The fruit of this plant turns bright red at maturity and resembles an edible mango fruit. Several reports are available in Southern India and Eastern Sri Lanka indicating the use of *Cerbera manghas* fruits for self-poisoning. Cerberin, neriifolin, and cerberoside are some of the main chemical constituents in this plant. The toxins could inhibit Na^+/K^+ ATPase and may result in vomiting, cardiac dysrhythmias, and hyperkalemia [32].

9.3.5 *Thevetia peruviana* (yellow oleander)

Thevetia peruviana is cultivated as an ornamental tree and is commonly found in the tropics and subtropics (Figure 9.1). The seeds contain high concentrations of toxic cardiac glycosides like thevetin A, thevetin B, and neriifolin. Accidental poisoning occurs among children and adults due to the consumption of yellow oleander leaves in herbal teas. The deliberate ingestion of seeds of this plant is a popular method of self-harm among South Asians. Vomiting, diarrhea, bradycardia, cardiac dysrhythmias, and death can occur due to the poisoning [33–35].

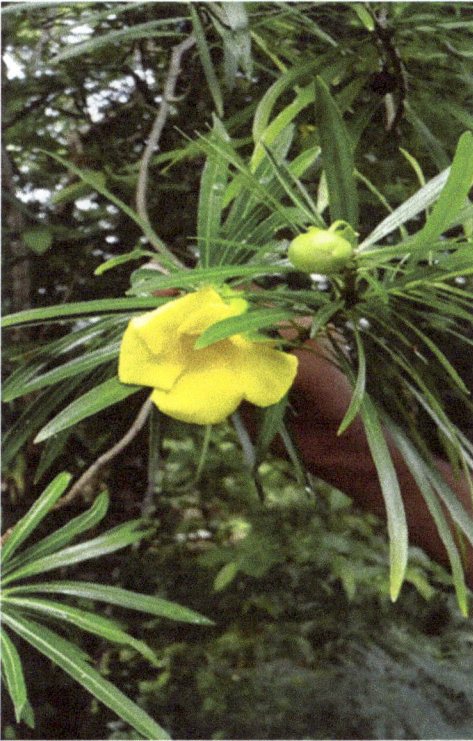

Figure 9.1: *Thevetia peruviana.*

9.3.6 *Nerium oleander* (oleander)

Nerium oleander is an ornamental tree with pink-, red-, white-, peach-, or yellow-colored flowers. Cardiac glycosides are abundant in all parts of the plant. Oleandrin and neriin are the most potent cardiac glycosides which may exert inhibitory effects on the Na^+/K^+ ATPase pump in the cardiac tissues. As a result, cardiac and gastrointestinal symptoms could be observed four hours after the ingestion [36].

The attractiveness of the flower could be a reason for accidental ingestion of parts of the plant by small children. As the plant is used in folk medicine, overdosing can occur while it is commonly used in homicides or suicides [36, 37].

9.3.7 *Gloriosa superba* (flame lily, climbing lily)

Despite the wide utility in folk medicine, *Gloriosa superba* is reputed as a highly poisonous plant (Figure 9.2). Every part of the plant is poisonous; however, the toxic effects are more conspicuous in tubers. Several alkaloids are abundant in this plant, of which colchicine and gloriosine are mainly responsible for fatal complications associated with *G. superba* poisoning. Colchicine and gloriosine have direct effects on rapidly proliferating cells due to the anti-mitotic activity. As a result, mitosis may arrest in metaphase of the cell cycle. Neurological, cardiac, gastrointestinal, and bone marrow toxicity could be observed within a few hours of the ingestion [38].

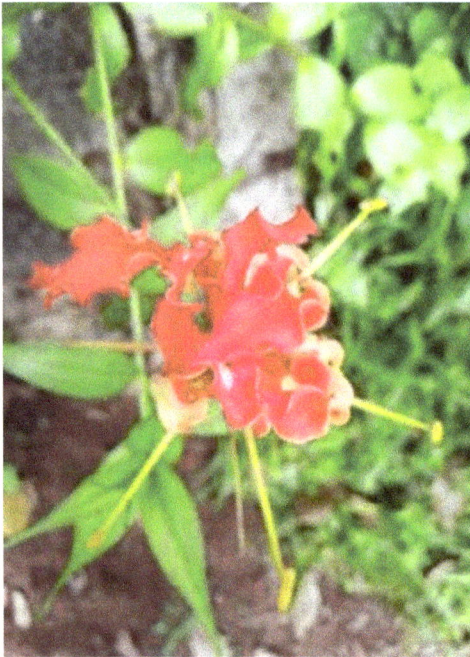

Figure 9.2: *Gloriosa superba.*

9.3.8 *Jatropha curcas* (physic nut/purging nut/Barbados nut)

Jatropha curcas is a traditional medicinal plant in South Asian countries and as of recent times, it is being cultivated as a biodiesel fuel. Accidental ingestion of seeds of this plant is prevalent among children in countries like India. The presence of curcin, ricin, and cyanic acid is responsible for the toxic effects of this plant. Every part of the plant is poisonous; however, seeds contain the highest concentration of ricin. Accidental poisoning is common in children and gastrointestinal complications like vomiting, diarrhea, and abdominal pain may occur within a short period following the ingestion [39, 40].

9.3.9 *Ricinus communis* (castor oil plant)

Ricinus communis is used in traditional medicine in many countries for various purposes, for example, as a laxative, to treat infections and inflammation, etc. Apart from medical applications, castor seed oil is used as a fuel for oil lamps [41].

The toxalbumin called ricin is the main toxic principle of this plant. It is a protein macromolecule comprised of two polypeptide chains – chain A and B. These two chains are held together by a single disulfide bond. Chain B binds to the cell surface facilitating the entry of the toxin to the cell, whereas chain A activates the 60S ribosomal subunit and thereby disrupts protein synthesis [42].

The toxic effects could be seen only when the outer shell of the seed is broken or chewed. Therefore the ingestion of castor beans is not always associated with toxic effects [43].

9.3.10 *Jatropha multifida*

Like other species in the genus *Jatropha, Jatropha multifida* also contains ricin [42]. Some reports indicate that this plant contains jatrophin that may result in agglutination and hemolysis of red blood cells [44].

9.3.11 *Adenia palmata*

The tuber and the fruit of *Adenia palmata* are used in folk medicine as a cure for snake bites. This plant has a close resemblance to passion fruit (*Passiflora edulis*); thus, it is quite difficult to distinguish between the two. Therefore, the morphological similarity between the two plants is accountable for the accidental ingestion of *A. palmata* [45].

9.3.12 *Abrus precatorius* (rosary pea)

Abrus precatorius (Figure 9.3) is used in traditional medicine, although ingestion of the seeds becomes a popular method of self-harm. Every part of this plant is poisonous, and the toxic principle has been identified as toxalbumin abrin which can inhibit protein synthesis in eukaryotic cells. The intact seeds may pass through the gastrointestinal tract without causing any toxicity, while the ingestion of seeds that were damaged can cause acute hemorrhagic gastroenteritis [46, 47].

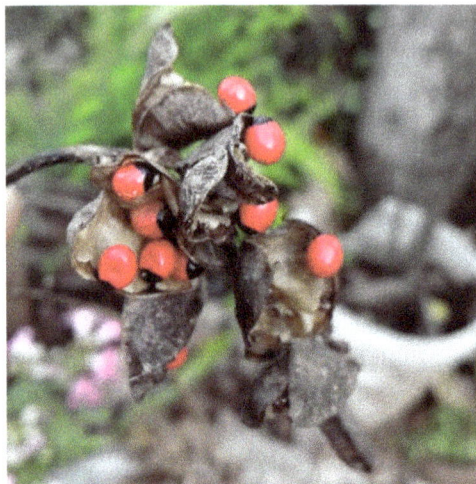

Figure 9.3: *Abrus precatorius.*

9.3.13 *Manihot esculenta* (cassava/manioc)

Cassava is considered as an important source of calories in many countries, specially in Africa, Latin America, and Asia. However, there are numerous reports on acute poisoning associated with the consumption of cassava-based meals. The symptoms may vary from dizziness, nausea, abdominal pain, and diarrhea, while death could also happen [48, 49].

Several cyanogenic glycosides are abundant in every part of the plant; however, leaves are more toxic than the roots. Linamarin and lotaustralin are the most prevalent cyanogenic glycosides in cassava. These cyanogenic glycosides can undergo acid, enzymatic, or thermal hydrolysis and release HCN [48]. However peeling, prolonged soaking and thorough cooking of cassava roots can effectively reduce the cyanide content [49].

9.3.14 *Alocasia macrorrhiza* (giant elephant's ear)

Alocasia macrorrhiza is a household plant (Figure 9.4) and is also used in traditional medicine to treat conditions like influenza, high fever, typhoid fever, rheumatic fever, and pulmonary tuberculosis. The tuber of this plant contains alomacrorrhiza A and allocasin. Calcium oxalate is distributed in the entire plant, and it is believed to be responsible for the development of ulcerative lesions while the presence of sapotoxin can include gastroenteritis and paralysis of the nerve centers following the ingestion of the plant [50, 51].

Figure 9.4: *Alocasia macrorrhiza.*

9.3.15 *Dieffenbachia amoena* (dumb cane)

Dieffenbachia amoena is cultivated as an ornamental plant. *Dieffenbachia* spp. contain calcium oxalate crystals as well as toxic proteins in all parts including sap. Accidental ingestion is associated with intense pain, inflammation of mouth and throat, anorexia, and diarrhea [52].

9.4 Impact of poisonous plants on companion animals and livestock industry

Toxic plants are accountable for many incidences of poisoning in farm and companion animals. Accidental ingestion of indoor or garden plants is common among pets and companion animals, while contamination of hay/fodder with poisonous plants may exert life-threatening effects on livestock and horses [53].

In addition to plants well known for the presence of toxic metabolites like oleander (*N. oleander*), castor bean (*R. communis*), autumn crocus (*Colchicum autumnale*), and sago palm (*Cycas* spp.), several other plant species, for example, lily (*Lilium* and *Hemerocallis* spp.), azalea (*Rhododendron* spp.), and Kalanchoe (*Kalanchoe* spp.), are identified as poisonous plants causing serious systemic effects on household pets, specially cats and dogs [54]. Poisoning incidences are generally underdiagnosed or go unnoticed due to the occurrence of non-specific clinical features, difficulties in spotting the ingestion of poisonous plant materials by animals, and lack of knowledge on toxic plants [55]. Some plant species accountable for companion animal poisoning incidences are summarized in Table 9.1 [55].

Table 9.1: Toxic plants responsible for the poisoning of companion animals.

Plant	Toxic compound/s	Toxicity manifestation
Aucuba japonica	Aucubin	Mild diarrhea, vomiting
Cycas revoluta	Macrozamin, neocycasin, cycasin	Diarrhea, liver damage, hypoproteinemia, hypoglycemia, thrombocytopenia
Cyclamen spp.	Saxifragifolin B, cyclamin	Sialorrhea, gastrointestinal symptoms heart rhythm abnormalities, seizures
Dieffenbachia spp.	Insoluble calcium oxalates, trypsin-like protease	Sialorrhea, dysphagia vomiting, diarrhea, keratoconjunctivitis, corneal ulceration, eyelids edema
Dracaena marginata	Steroidal saponins	Hypersalivation, gastrointestinal signs, weakness, incoordination, and mydriasis
Ficus benjamina	Ficin, furocoumarins, ficusin	Gastrointestinal symptoms, dermal irritation
Euphorbia pulcherrima	Diterpenoid euphorbol esters, steroids	Vesicular dermatitis, conjunctivitis, stomatitis, vomiting, diarrhea
Lilium spp.	Steroidal glycoalkaloids, steroidal saponins	Anorexia, lethargy, gastrointestinal signs, sialorrhea, acute renal failure due to tubular necrosis
Rhododendron spp.	Grayanotoxin I and other grayanotoxin glycosides	Increased vagal tone with vomiting, diarrhea or constipation, dyspnea, cardiac alterations (tachycardia or bradycardia, arrhythmia, hypotension, and collapse), paralysis, convulsions

Table 9.1 (continued)

Plant	Toxic compound/s	Toxicity manifestation
Spathiphyllum spp.	Insoluble calcium oxalates	Oral irritation/burning, drooling, dysphagia, vomiting
Zantedeschia aethiopica	Insoluble calcium oxalates crystals (raphides)	Oral hyperemia and edema, hypersalivation, anorexia, depression. gastrointestinal signs (vomiting, diarrhea, abdominal pain), dermatitis
Nandina domestica	Cyanogenic glycoside	Vomiting, dyspnea, cherry-red mucous membranes, respiratory failure, convulsions.

Moreover, poisonous plants cause significant losses to the livestock industry. The livestock operator should be well aware of the poisonous plants grown in the area/ pasture and how to control or avoid poisoning. Some plants are extremely toxic; therefore, ingestion of very small amounts may result in life-threatening conditions. Other plant species may cause poisoning effects only after exposure to toxic plant material over weeks or months, and at this stage, the treatment may not be very effective [56].

There are numerous reports of poisoning incidences. In 1995, 72 out of 476 cattle in the Dar es Salaam area of Tanzania died after being offered with hey mixed with the poisonous plant *Dichapetalum* [57]. Similarly, 50 lactating Fleckvieh cows were affected after consuming fodder accidentally contaminated with dry oleander pruning wastes [58]. Furthermore, Mendonça et al. [59] reported that cattle in the semiarid region of Pernambuco, Brazil had developed gastrointestinal, cardiovascular, and neuromuscular disorders due to the ingestion of *Kalanchoe* spp. that contain cardiotoxic glycosides [59]. Another report revealed the development of severe acute gastrointestinal irritation in a heifer after ingesting fresh leaves of *Colchicum autumnale* on a damp meadow [60].

It is generally noticed these animals do not eat poisonous plants except when they are forced to do so by hunger. Therefore by adopting proper pasture management practices to provide ample forage and thereby encouraging the consumption of non-toxic plants, many of these poisoning incidences can be prevented [56].

9.5 Conclusion

Plants produce a diverse array of secondary metabolites for defense purposes, which may exert toxic effects on humans, pets, and companion animals as well as livestock animals. Accidental or intentional contamination/ingestion of some plant species can lead to toxic and/or fatal reactions. Many of these poisoning incidences can be prevented by increasing public awareness on the identification of poisonous plants and their toxic effects.

References

[1] Wink M. Plant breeding: Importance of plant secondary metabolites for protection against pathogens and herbivores. Theor Appl Genet 1988, 75, 225–233.

[2] Wink M. Evolution of secondary metabolites from an ecological and molecular phylogenetic perspective. Phytochemistry 2003, 64, 3–19.

[3] Wink M. Plant secondary metabolites modulate insect behavior-steps toward addiction?. Front Physiol 2018, 9, 364, 10.3389/fphys.2018.00364.

[4] Wink M. Mode of action and toxicology of plant toxins and poisonous plants. Mitt Julius Kühn-Inst 2009, 421, 93–112.

[5] Botha CJ, Penrith ML. Poisonous plants of veterinary and human importance in southern Africa. J Ethnopharmacol 2008, 119(3), 549–558.

[6] Anywar G. 2020. Historical use of toxic plants. In: Mtew AG, Egbuna C, Rao GMN, eds, Poisonous Plants and Phytochemicals in Drug Discovery. 1st. NJ, USA, John Wiley & Sons, 2021, 1–17.

[7] Leopold AC, Ardrey R. Toxic substances in plants and the food habits of early man. Science 1972, 176(4034), 512–514.

[8] Raghavendra T. Neuromuscular blocking drugs: Discovery and development. J R Soc Med 2002, 95(7), 363–367.

[9] Doherty JE, Kane JJ. Clinical pharmacology and therapeutic use of digitalis glycosides 1. Drugs 1973, 6, 182–221.

[10] Man S, Gao W, Wei C, Liu C. Anticancer drugs from traditional toxic Chinese medicines. Phytother Res 2012, 26(10), 1449–1465.

[11] Alkadi H, Khubeiz MJ, Jbeily R. Colchicine: A review on chemical structure and clinical usage. Infect Disord Drug Targets 2018, 18(2), 105–121.

[12] Hotti H, Rischer H. The killer of Socrates: Coniine and related alkaloids in the plant kingdom. Molecules 2017, 22(11), 1962. 10.3390/molecules22111962.

[13] Robb GL. The ordeal poisons of Madagascar and Africa. Bot Mus Leaf Harv Univ 1957, 17(10), 265–316.

[14] Tyagi N, Tyagi M, Pachauri M, Ghosh PC. Potential therapeutic applications of plant toxin-ricin in cancer: Challenges and advances. Tumour Biol 2015, 36(11), 8239–8246.

[15] Philippe G, Angenot L. Recent developments in the field of arrow and dart poisons. J Ethnopharmacol 2005, 100(1–2), 85–91.

[16] Neuwinger HD. Fish poisoning plants in Africa. Bot Acta 1994, 107(4), 263–270.

[17] Isman MB. Botanical insecticides, deterrents, and repellents in modern agriculture and an increasingly regulated world. Annu Rev Entomol 2006, 51, 45–66.

[18] Marrero E, Bulnes C, Sánchez LM, Palenzuela I, Stuart R, Jacobs F, Romero J. *Pteridium aquilinum* (bracken fern) toxicity in cattle in the humid Chaco of Tarija, Bolivia. Vet Hum Toxicol 2001, 43(3), 156–158.

[19] Arlt VM, Stiborova M, Schmeiser HH. Aristolochic acid as a probable human cancer hazard in herbal remedies: A review. Mutagenesis 2002, 17(4), 265–277.

[20] Gupta VK, Sharma B. Forensic applications of Indian traditional toxic plants and their constituents. Forensic Res Criminol Int J 2017, 4(1), 27–32.

[21] Dayasiri MBKC, Jayamanne SF, Jayasinghe CY. Plant poisoning among children in rural Sri Lanka. Int J Pediatr 2017, 2017, 6187487, 10.1155/2017/6187487.

[22] Vickery M. Plant poisons: Their occurrence, biochemistry and physiological properties. Sci Prog 2010, 93(2), 181–221.

[23] Eddleston M, Haggalla S. Fatal injury in Eastern Sri Lanka, with special reference to cardenolide self-poisoning with *Cerbera manghas* fruits. Clin Toxicol (Phila) 2008, 46(8), 745–748.

[24] Krenzelok EP. Aspects of *Datura* poisoning and treatment. Clin Toxicol 2010, 48(2), 104–110.

[25] Disel NR, Yilmaz M, Kekec Z, Karanlik M. Poisoned after dinner: Dolma with *Datura Stramonium*. Turk J Emerg Med 2016, 15(1), 51–55.

[26] Bouziri A, Hamdi A, Borgi A, Hadj SB, Fitouri Z, Menif K, Ben Jaballah N. *Datura stramonium* L. poisoning in a geophagous child: A case report. Int J Emerg Med 2011, 4(1), 31. 10.1186/1865-1380-4-31.

[27] Perera P, Samuelsson G, van Beek TA, Verpoorte R. Tertiary indole alkaloids from leaves of *Tabernaemontana dichotoma*. Planta Med 1983, 47(3), 148–150.

[28] Guo R, Wang T, Zhou G, Xu M, Yu X, Zhang X, Sui F, Li C, Tang L, Wang Z. Botany, Phytochemistry, pharmacology and toxicity of *Strychnos nux-vomica* L.: A review. Am J Chin Med 2018, 46(1), 1–23.

[29] Ponraj L, Mishra AK, Koshy M, Carey RAB. A rare case report of *Strychnos nux-vomica* poisoning with bradycardia. J Family Med Prim Care 2017, 6(3), 663–665.

[30] Maji AK, Banerji P. *Strychnos nux-vomica*: A poisonous plant with various aspects of therapeutic significance. J Basic Clin Pharma 2017, 8, S087–S103.

[31] Tsai YC, Chen CY, Yang NI, Yang CC. Cardiac glycoside poisoning following suicidal ingestion of *Cerbera manghas*. Clin Toxicol (Phila) 2008, 46(4), 340–341.

[32] Selladurai P, Thadsanamoorthy S, Ariaranee G. Epidemic self-poisoning with seeds of *Cerbera manghas* in Eastern Sri Lanka: An analysis of admissions and outcome. J Clin Toxicol 2016, 6(2), 287. 10.4172/2161-0495.1000287.

[33] Eddleston M, Ariaratnam CA, Sjöström L, Jayalath S, Rajakanthan K, Rajapakse S, Colbert D, Meyer WP, Perera G, Attapattu S, Kularatne SAM, Sheriff MR, Warrell DA. Acute yellow oleander (*Thevetia peruviana*) poisoning: Cardiac arrhythmias, electrolyte disturbances, and serum cardiac glycoside concentrations on presentation to hospital. Heart 2000, 83, 301–306.

[34] Eddleston M, Ariaratnam CA, Meyer WP, Perera G, Kularatne AM, Attapattu S, Sheriff MHR, Warrell DA. (1999), Epidemic of self-poisoning with seeds of the yellow oleander tree (*Thevetia peruviana*) in Northern Sri Lanka. Trop Med Int Health 1999, 4, 266–273.

[35] Rajapakse S. Management of yellow oleander poisoning. Clin Toxicol (Phila) 2009, 47(3), 206–212.

[36] Khan I, Kant C, Sanwaria A, Meena L. Acute cardiac toxicity of *Nerium oleander/indicum* poisoning (kaner) poisoning. Heart Views 2010, 11(3), 115–116. 10.4103/1995-705X.76803.

[37] Farkhondeh T, Kianmehr M, Kazemi T, Samarghandian S, Khazdair MR. Toxicity effects of *Nerium oleander*, basic and clinical evidence: A comprehensive review. Hum Exp Toxicol 2020, 39(6), 773–784.

[38] Premaratna R, Weerasinghe MS, Premawardana NP, de Silva HJ. *Gloriosa superba* poisoning mimicking an acute infection- a case report. BMC Pharmacol Toxicol 2015, 16, 27, 10.1186/s40360-015-0029-6.

[39] Singh RK, Singh D, Mahendrakar AG. *Jatropha* poisoning in children. Med J Armed Forces India 2010, 66(1), 80–81.

[40] Gupta A, Kumar A, Agarwal A, Osawa M, Verma A. Acute accidental mass poisoning by *Jatropha curcas* in Agra, North India. Egypt J Forensic Sci 2016, 6(4), 496–500.

[41] Worbs S, Köhler K, Pauly D, Avondet MA, Schaer M, Dorner MB, Dorner BG. *Ricinus communis* intoxications in human and veterinary medicine-a summary of real cases. Toxins (Basel) 2011, 3(10), 1332–1372.

[42] Levin Y, Sherer Y, Bibi H, Schlesinger M, Hay E. Rare *Jatropha multifida* intoxication in two children. J Emerg Med 2000, 19(2), 173–175.

[43] Al-Tamimi FA, Hegazi AE. A case of castor bean poisoning. Sultan Qaboos Univ Med J 2008, 8(1), 83–87.

[44] Guruge K, Seneviratne A, Badureliya C. A case of *Jatropha multifida* poisoning. Sri Lanka J Child Health 2008, 36(4), 148.

[45] Ratnatilaka A, Yakandawala D, Ratnayake J, Sugathadasa S. Poisoning with "hondala" leaves due to misidentification as "passion fruit" leaves. Ceylon Med J 2011, 48(1), 23.

[46] Subrahmanyan D, Mathew J, Raj M. An unusual manifestation of *Abrus precatorius* poisoning: A report of two cases. Clin Toxicol (Phila) 2008, 46(2), 173–175.

[47] Huang J, Zhang W, Li X, Feng S, Ye G, Wei H, Gong X. Acute abrin poisoning treated with continuous renal replacement therapy and hemoperfusion successfully. Medicine 2017, 96(27), e7423. 10.1097/md.0000000000007423.

[48] Teles FFF. Chronic poisoning by hydrogen cyanide in cassava and its prevention in Africa and Latin America. Food Nutr Bull 2002, 23(4), 407–412.

[49] Peñas J, de Los Reyes VC, Sucaldito MN, Manalili DL, Hizon H, Magpantay R. A retrospective cohort study on cassava food poisoning, Santa Cruz, Davao del Sur, Philippines, October 2015. Western Pac Surveill Response J 2018, 9(4), 7–11.

[50] Joshi A, Karnawat BS, Narayan JP, Sharma V. *Alocasia macrorrhiza*: A decorative but dangerous plant. Int J Sci Stud 2015, 3(1), 221–223.

[51] Lin TJ, Hung DZ, Hu WH, Yang DY, Wu TC, Deng JF. Calcium oxalate is the main toxic component in clinical presentations of *Alocasia macrorrhiza* (L) Schott and Endl poisonings. Vet Hum Toxicol 1998, 40(2), 93–95.

[52] Chirayath D, Ambily VR, Amrutha VS. *Dieffenbachia amoena* poisoning in a goat – a case report. JIVA 2017, 15(3), 43–44.

[53] Cortinovis C, Caloni F. Plants toxic to farm and companion animals. In: Gopalakrishnakone P, Carlini C, Ligabue-Braun R, eds,Plant Toxins- Toxinology. 1st, Dordrecht, NL, Springer, 2017, 107–134.

[54] Milewski LM, Khan SA. (2006). An overview of potentially life-threatening poisonous plants in dogs and cats. J Vet Emerg Crit Care 2006, 16, 25–33.

[55] Alessia B, Paola F, Francesca C. Indoor companion animal poisoning by plants in Europe. Front Vet Sci 2020, 7, 487, 10.3389/fvets.2020.00487.

[56] Forero L, Nader G, Craigmill A, Ditomaso JM, Puschner B, Maas J Livestock-poisoning plants of California. 2011. http://dx.doi.org/10.3733/ucanr.8398 Retrieved from https://escholarship.org/uc/item/0h1766z0

[57] Msami HM. An outbreak of suspected poisoning of cattle by *Dichapetalum* sp. Tanzania Trop Anim Health Prod 1999, 31(1), 1–7.

[58] Ceci L, Girolami F, Capucchio MT, Colombino E, Nebbia C, Gosetti F, Marengo E, Iarussi F, Carelli G. Outbreak of oleander (*Nerium oleander*) poisoning in dairy cattle: Clinical and food safety implications. Toxins (Basel) 2020, 12(8), 471. 10.3390/toxins12080471.
[59] Mendonça FS, Nascimento NCF, Almeida VM, Braga TC, Ribeiro DP, Chaves HAS, Silva Filho GB, Riet-Correa F. An outbreak of poisoning by *Kalanchoe blossfeldiana* in cattle in northeastern Brazil. Trop Anim Health Prod 2018, 50(3), 693–696.
[60] Kupper J, Rentsch K, Mittelholzer A, Artho R, Meyer S, Kupferschmidt H, Naegeli H. A fatal case of autumn crocus (*Colchicum autumnale*) poisoning in a heifer: Confirmation by mass-spectrometric colchicine detection. J Vet Diagn Invest 2010, 22(1), 119–122.

Mayuri Napagoda

10 Quality and safety of herbal medicinal products

10.1 Introduction

The World Health Organization defines herbal medicines as plant-derived materials or preparations intended for human therapeutic use or for other health benefits in humans [1, 2]. Nevertheless, most of the traditional herbal formulations also contain animal material and/or mineral compounds. Herbal products are consumed raw as tea or as concentrated extracts (decoctions), applied as a paste or powder on skin, or sometimes available as pills or liquids. The global trend in returning to natural or alternative therapies has dramatically increased the use of plant based-medicines and other botanicals over the past few years. A large number of botanicals have been transformed into various pharmaceutical forms like tablets, capsules, powders, and syrups, while some have been developed into cosmeceuticals, fragrances, dietary supplements, and nutraceuticals. Although plant based-medicines are traditionally considered harmless and thus extensively being utilized by people without prescriptions, the number of reports on health issues associated with herbal medicines could not be neglected. The presence of contaminants or adulterants, as well as inherent toxicity of plant ingredients, may cause direct toxic effects, interactions with other drugs, or reduce the efficacy of herbal formulations. Therefore, assessment of the safety and standardization of herbal formulations is one of the priority areas in herbal medicinal research, and it requires comprehensive phytochemical and pharmacological studies [3, 4].

10.2 Contamination and adulteration of herbal medicinal products

Adulteration of herbal medicinal products is defined as "fraudulent practices in which a herbal medicinal product is substituted partially or fully with impure, extraneous, improper or inferior products/substances." On the other hand, contamination is referred to as "the undesired introduction of impurities of a chemical or microbiological nature, or of foreign matter, into or onto a starting material, intermediate product or finished herbal product during production, sampling, packaging or repackaging, storage or transport" [5].

Mayuri Napagoda, Faculty of Medicine, University of Ruhuna, 80000, Galle, Sri Lanka,
e-mail: mayurinapagoda@yahoo.com

https://doi.org/10.1515/9783110595949-010

10.2.1 Contaminants

Microorganisms and the microbial toxins, pesticides and fumigation residues, dust, pollen, and heavy metals are some potential contaminants of herbal medicinal products (Table 10.1).

Table 10.1: Major contaminants in herbal medicines [6].

	Type of contaminant	Examples
1.	Microorganisms	
	Bacteria	*Staphylococcus aureus*, *Pseudomonas aeruginosa*, *Salmonella* spp., *Shigella* spp., *Escherichia coli*
	Fungi	Yeast, molds
2.	Microbial toxins	Mycotoxins, Bacterial endotoxins
3.	Animals	
	Parasites	Protozoa – amoebae Helminths –Nematoda
	Insects and others	Cockroach and its parts, mouse excreta, earthworms
4.	Toxic metals and nonmetals	Lead, cadmium, mercury, chromium, arsenic, nitrite
5.	Agrochemical residues	
	Insecticides	Carbamate, chlorinated hydrocarbons, organophosphorus
	Herbicides	2,4-Dichlorophenoxyacetic acid, 2,4,5-Trichlorophenoxyacetic acid
	Fungicides	Dithiocarbamate
	Fumigants	Ethylene oxide, phosphine, methyl bromide, sulfur dioxide
	Antiviral agents	Thiamethoxam
6.	Residual solvents	Acetone, methanol, ethanol, butanol
7.	Radioactivity	Cs-134, Cs-137

The unscientific methods adopted during harvesting, handling, storage, and transportation of raw materials as well as humid climatic conditions can trigger microbial infestations in herbal medicines [7]. The microorganisms often isolated from herbal medicines include *Staphylococcus aureus*, *Escherichia coli*, *Salmonella*, *Shigella*, *Pseudomonas aeruginosa*, and fungi like *Aspergillus*. For example, de Sousa Lima et al. [8]

observed bacterial and fungal growth in 51.5% and 35.6% of herbal medicinal samples, respectively, where *S. aureus*, *Salmonella* spp., *E. coli*, and *P. aeruginosa* have been identified as the most common microbial contaminants in these samples. The levels of viable bacteria and fungi were found to be above the safety levels in some instances [8]. Similarly, Yesuf et al. [9] reported the presence of *Bacillus* spp., *Enterobacter* spp., *Shigella dysenteriae*, and *Salmonella* spp. in herbal medicinal preparations collected from Gondar Town, Ethiopia [9]. Moreover, Kalumbi et al. [10] found that the herbal medicinal formulations obtained from different markets in Blantyre, Malawi, were contaminated with *Bacillus*, coagulase-negative *Staphylococcus*, *Klebsiella* spp., *Enterobacter* spp., etc. and also with heavy metals, lead and cadmium. Further, they discovered that the level of microbial and lead contamination was far above the regulatory limits [10]. Using four Chinese herbal medicinal formulations that are popular in Malaysia, Ting et al. [11] demonstrated that the boiling process involved in the preparation of decoctions can significantly reduce microbial and heavy metal contaminants [11].

The presence of mycotoxins in herbal products has frequently been reported in many countries. Mycotoxins are a group of fungal secondary metabolites that can cause carcinogenic or other toxic effects on humans and animals. *Aspergillus*, *Penicillium*, *Fusarium*, and *Alternaria* are some examples of toxigenic fungal genera, while aflatoxins, ochratoxin A, fumonisins, zearalenone, and deoxynivalenol have been identified as the most harmful fungal toxins [12]. In a study conducted in India, 858 fungal isolates including *Aspergillus flavus* that produces aflatoxin B1 were detected in samples prepared from *Adhatoda vasica*, *Asparagus racemosus*, *Evolvulus alsinoides*, *Glycyrrhiza glabra*, *Plumbago zeylanica*, and *Terminalia chebula*. Interestingly, the essential oil of *Cinnamomum camphora* was capable of inhibiting the growth of these toxigenic *A. flavus* and thus the production of aflatoxin B1 [13]. Moreover, aflatoxins and/or ochratoxin A were detected in many medicinal herbs (ginseng, ginger, liquorice, turmeric, kava-kava, etc.) and in dried fruits (figs, apricots, plums, dates, etc.) [14].

Herbal medicinal preparations sometimes get contaminated with toxic heavy metals, imposing serious health risks to consumers [15]. Lead, arsenic, mercury, cadmium, nickel, and chromium are the most common heavy metals detected in raw materials used for herbal remedies [16]. As plants are capable of accumulating heavy metals, consumption of herbal products formulated using medicinal plants grown in polluted sites can lead to severe consequences on human health. Moreover, medicinal plants are at a high risk of being contaminated with heavy metals due to the increased anthropogenic activities like mining, irrigation with contaminated water, and the use of pesticides and fertilizers [17]. A study conducted on the heavy metal content in some traditional herbal remedies available in the USA, Vietnam, and China revealed that the arsenic, lead, and mercury concentrations in some products had exceeded the levels recommended in public health guidelines [18].

Agrochemical residues are also detectable in herbal drugs. In order to intensify the cultivation and to obtain high yields of good quality products, different agricultural practices like the use of fertilizers and pesticides, and administering fumigants during storage and transportation are adopted by the farmers. These practices often result in the contamination of herbal products with agrochemical residues [19–21]. Synthetic pesticides belonging to organochlorine, organophosphorus, and pyrethroid groups are most widely reported in herbal drugs [22]; for example, Opuni et al. [23] detected the contamination of herbal medicinal products with chlorpyrifos (an organophosphate) and/or bifenthrin (a pyrethroid) [23]. In another study, residues of 16 agrochemicals were detected in 72.1% of the analyzed herbal materials where azoxystrobin (fungicide), linuron (herbicide), carbendazim (fungicide), metalaxyl, and metalaxyl M (fungicides) and dimethoate (insecticide and acaricide) were most frequently present [20].

Apart from the aforementioned contaminants, medicinal herbs, herbal preparations, or products can be subjected to cross-contamination from extraneous materials such as dust, plastics, glass, radionuclide, and other materials.

10.2.2 Adulteration

Adulteration of herbal drugs is a major issue in the herbal industry. This involves many practices like substitution of one or more ingredient(s) with substandard cheap ingredients, substitution with artificially manufactured substances, substitution with non-drug components, and the addition of foreign non-drug materials [24].

Ginseng is an internationally popular herbal medicine that belongs to the genus *Panax*. There are about 14 species, out of which *Panax ginseng* (Asian ginseng) and *Panax quinquefolium* (American ginseng) are widely used in medicine and contain various bioactive compounds, particularly different types of ginsenosides. According to traditional Chinese medicine, Asian ginseng is considered "hot"; therefore, only a few people can use it. On the contrary, American ginseng is "cool"; thus, most people can use it. Moreover, American ginseng is reported to be more effective than its Asian counterpart and as a result, roots of *Panax quinquefolium* are usually five to ten times more expensive than the roots of *Panax ginseng*. With the increasing market demand and profit temptation, *Panax quinquefolium* has been subjected to substitution and/or adulteration in the commercial market [25–27]. Moreover, there are instances where *Codonopsis lanceolata* (deodeok) and *Platycodon grandiflorum* (doraji) are being marketed as original ginseng [27, 28].

Several incidents of serious intoxications have been caused by adulterants or substitutes. Adulteration can be resulted due to the misidentification of plant materials and confusion of species, as well as inappropriate labeling of raw materials [29].

Similar common names in different plant species have resulted in the misidentification of plant materials. A well-known example is a substitution of non-toxic *Stephania tetrandra* (Fang Ji) with *Aristolochia fangchi* (Guang Fang Ji), a plant that contains nephrotoxic and carcinogenic aristolochic acid [30]. More than 70 cases of renal failure were reported in Belgium after consumption of a weight-loss preparation in which *S. tetrandra* was substituted with *A. fangchi* because of their similar names [3].

Two cases of encephalopathy and neuropathy were reported in Hong Kong in 1989 following ingestion of a decoction containing *Podopyllum hexandrum*. *P. hexandrum* contains the neurotoxin podophyllotoxin in high concentration; nevertheless, it is used as an adulterant of "long dan cao" (*Gentiana* spp). Around the same period in Taipei and Kuala Lumpur, the erroneous substitution of *P. hexandrum* for Wai-Ling-Sin (*Clematis chinensis*) had caused several cases of neuropathy [31, 32].

Both *Solanum lyratum* and *Aristolochia mollissima* share the common Chinese name "Bai Mao Teng," although these two plants are belonging to two different families. The confusion over the name had led to serious consequences. *S. lyratum* is not harmful; however, *A. mollissima* contains aristolochic acid, a phytochemical with nephrotoxic and carcinogenic properties [29, 33]. Another example of confusion over the common name is the Chinese diuretic drug "Mu Tong." According to the classical Chinese herbal literature, until the mid-seventeeth century, several *Akebia* species were utilized as the original source plants of Mu Tong. Thereafter, *Clematis* species were recognized as the main source of Mu Tong. However, since the 1950s *Aristolochia manshuriensis* is considered as the source of the Chinese herb Mu Tong, and this had led to serious confusion as the substitution with *A. manshuriensis* may cause renal failure [29, 34].

The adulteration of herbal medicine with undeclared synthetic drugs to enhance their therapeutic effect has been reported in many countries. A study conducted in Taiwan revealed that 23.7% of herbal samples were adulterated with pharmaceuticals [35], while 7% of proprietary herbal products in California contained undeclared pharmaceuticals [36]. Similarly, a study conducted in Iran revealed the existence of illegal adulterants in 63 herbal weight loss formulations. Those samples mainly contained sibutramine beyond its therapeutic dose posing serious risks to the health of the people who consume the herbal formulation [37]. In another study conducted in Oman, sildenafil, tadalafil and vardenafil were detected as adulterants in several herbal medicinal products [38]. Moreover, paracetamol, dexamethasone, and prednisolone were identified in pain reliever formulations [39] while aminopyrine, indomethacin, hydrocortisone phenylbutazone, etc. were detected in several Chinese herbal medicines [40].

10.3 Safety concerns on herbal medicines

Many people believe that herbal medicines with a long history of popular use are normally safe at their therapeutic dose. However, the absence of records of adverse effects may not always be an indicator of lack of toxicity specially concerning the long-term adverse effects that are difficult to detect. Therefore a properly designed epidemiology study (preferably, a prospective cohort study), as well as comprehensive pre-clinical and clinical studies, is required to determine the safety of popular herbal medicines [41].

The inappropriate usage of herbal medicines can lead to adverse effects and toxicities. Sometimes, these products are used for inappropriate indications (non-traditional indications like weight loss, athletic performance, recreational use), or prolonged periods or in inappropriate dosages. For example, in traditional Chinese medicine, *Ephedra* is used in small doses as a remedy for wheezing and cough but is never recommended as a stimulant, a dieting agent, or a recreational agent [2]. However, *Ephedra* was an ingredient of a multicomponent dietary supplement generally popular as a weight loss and an energy enhancement agent. The Food and Drug Administration had to ban the sale of this supplement in 2004 due to serious cardiovascular adverse effects associated with its use in excessive doses and durations [42]. Similarly, fatalities were reported after the intentional ingestion of seeds of *Datura stramonium* for recreational purposes rather than for its therapeutic effects [43].

Inappropriate processing is also an underlying reason for toxic effects; thus, herbs should be processed by adopting recommended protocols. An example is the processing of dried seeds of *Strychnos nux-vomica* that is used in traditional Chinese medicine to promote blood circulation, alleviate blood stasis, and relieve pain. The secondary metabolites strychnine and brucine, which possess strong convulsant action, are present in the seeds and as a result, it is recommended to process seeds by parching in a sand bath or frying in an oil bath at 235 °C. This type of processing can decompose or transform the above secondary metabolites into less toxic forms [39]. Similarly, *Aconitum* species are considered as an important drug in Traditional Chinese Medicine despite the presence of the toxic alkaloid aconitine and other cardiotoxic and neurotoxic metabolites like mesaconitine and hypaconitine. The traditional processing method of *Aconitum* roots is known as "Paozhi," and this process can reduce the toxicity of the plant material by degrading the diester diterpene alkaloids to the less toxic monoester diterpene alkaloids. Another example is the improper usage of *Aconitum* roots that resulted in several fatalities, particularly after drinking homemade medicated liquor containing *Aconitum* [44].

Adulteration of herbal medicines with modern pharmaceuticals such as nonsteroidal anti-inflammatory drugs (NSAIDs), steroids, antihistamines, and sexual-enhancing drugs can lead to serious adverse effects such as allergic reactions, Addisonian crisis, Cushing's syndrome, fatal hypoglycemia, and even death [2].

Heavy metals exist as a regular and deliberate component in many Asian herbal medicinal formulations and have been included for a specific curative purpose [45]. Consumption of such products may sometimes lead to heavy metal poisoning; for example; anemia, abdominal pain, and encephalopathy are associated with lead poisoning, while arsenic poisoning can be presented as leucopenia, anemia, sensory neuropathy, and malignancies [2].

Patients who are on multiple medications, at extremes of age, and with chronic illnesses have a higher risk of developing harmful drug–herbs interactions due to the concurrent use of herbal medicinal products and modern pharmaceuticals [46]. One of the most documented examples is the anticoagulant drug warfarin–herb interactions. Most often patients on warfarin are maintained on long-term therapy and as a result, many of them move concurrently to herbal remedies to achieve synergistic effects which may lead to serious conditions like intracranial hematoma. Herbs like *Salvia miltiorrhiza, Ginkgo biloba, Angelica sinensis, Panax quinquefolius, Carthamus tinctorius, Prunus persica, Glycyrrhiza glabra, Panax ginseng, Lycium barbarum, Zingiber officinale*, and *Panax notoginseng* were found to exhibit strong interactions with warfarin [47]. Moreover, augmentation of sedative effects of modern pharmaceuticals in the presence of medicinal plants with sedative effects like *Piper methysticum* and *Valeriana officinalis* has been documented. The long-term usage of *P. methysticum* can cause hepatotoxicity and dermatopathy [2, 41].

Another problem associated with the concurrent use of herbal medicines and modern pharmaceuticals is the interaction with metabolic enzymes, specially CYP3A, the most important cytochrome P450 isoform which is responsible for the metabolism of many drugs. The extracts of *Hypericum perforatum* (St. John's wort), which is popular as an herbal anti-depressant, can strongly induce CYP3A. This can lead to either a sub-therapeutic effect in drugs that have been inactivated or toxic effects in drugs that are being activated by this group of enzymes [48]. Besides, St. John's wort can induce the expression of transmembrane transporter protein PgP in the liver and intestine. PgP is involved in the uptake and distribution of several clinically important drugs. St. John's wort may cause pharmacokinetic and pharmacodynamic interactions with drugs like cyclosporine, midazolam, oxycodone, tracolimus, digoxin, atorvastin, and verapamil. The secondary metabolite hyperforin functions as a potent inducer of CYP3A4 and PgP and the clinical evidence suggested that St John's wort extracts with low hyperforin content do not change the pharmacokinetics of drugs like cyclosporine and midazolam [2, 41, 49].

A large number of medicinal plants and phytochemicals thereof are suspected of being carcinogens, mutagens, or teratogens. Numerous reports indicated the acute toxicity, chronic toxicity, and genotoxicity in pyrrolizidine alkaloids present in plants like *Crotalaria, Heliotropium* and *Amsinckia* spp. Safrole is a major constituent in the essential oil of *Sassafras albidum* that has been widely used for medicinal and culinary purposes. Safrole is also found in *Piper betle* (betel quid), and *Areca catechu* (areca nut) commonly chewed in South and Southeast Asian countries because of their addictive psycho-stimulating effects. However, the regular consumption of betel quid and areca nut is one of the risk factors for the development of oral cancers. Moreover, the hepatocarcinogenic potential of the alkenylbenzenes known as asarones (α-, β-, and γ-asarone) was observed in many studies. Asarones have been isolated from a wide range of medicinal plants, including *Acorus calamus, Asarum europaeum*, and *Mosannona depressa* [50].

Some herbal medicines can cause severe hepatotoxic events. For example, the root of *Callilepis laureola* (impila) is extensively employed in traditional medicine in South Africa to treat stomach complaints, cough, tapeworm infections, and impotence. However, several cases of fatal poisoning due to the ingestion of impila were reported. The *in vitro* experiments revealed the impila-induced cytotoxicity in human hepatoblastoma Hep G2 cells due to depletion of cellular glutathione (GSH) [51]. Another example is *Larrea tridentata* (chaparral), a desert shrub grown in the Southwestern United States and Mexico and consumed by Native Americans against a variety of ailments. Although it has been introduced as a botanical dietary supplement, several clinical cases revealed the possible hepatotoxic effects associated with the ingestion of chaparral. Jaundice with a marked elevation in liver enzyme concentrations in serum was often observed in those patients. Thus it is believed that the consumption of chaparral may cause acute/chronic irreversible liver damage. Moreover, investigations have also revealed that renal and skin toxic effects are associated with the ingestion of chaparral [52]. Table 10.2 presents some plant species reported with carcinogenic, mutagenic, or hepatotoxic effects on human and/or rodent models.

Table 10.2: Carcinogenic, mutagenic, and hepatotoxic effects exhibited by some medicinal plants [41].

Plant	Chemical constituent	Traditional use	Toxicity
Aristolochia spp.	Aristolochic acid	In traditional Chinese medicine for arthritis, rheumatism, hepatitis	Nephrotoxicity, upper tract urothelial carcinoma
Symphytum officinale	Pyrrolizidine alkaloids	Traditional medicine in Africa, China, Ayurveda	Hepatotoxicity, hepatic venous occlusive disease, liver cancer, genotoxicity,
Euphorbia tirucalli	Phorbol esters	Traditional medicine in Africa.	Burkitt's lymphoma after co-exposure of *E. tirucalli* and Epstein Barr virus
Rubia tinctorum	Hydroxyanthraquinones, lucidin	In Ayurveda, and in Europe for kidney stones	Liver and kidney malignant tumors
Atractylis gummifera	Atractylosides, gummiferin	In Mediterranean region used as antipyretic, emetic, diuretic, chewing gum	Acute hepatitis, nephrotoxicity, hepatorenal failure
Chelidonium majus	Celandine	In Europe and temperate regions of Asia used to treat dyspepsia, biliary colic, cholelithiasis	Acute liver injury, moderate elevations of ALT, cholestasis

Table 10.2 (continued)

Plant	Chemical constituent	Traditional use	Toxicity
Teucrium chamaedrys and other spp.	Furanoditerpenoids	In Europe and Middle East	Hyperbilirubinemia, anorexia, nausea, marked elevations of ALT
Pteridium spp.	Sesquiterpenes and analogues; ptaquiloside	As food in East Asia and American Indians, food and traditional medicine in New Zealand (the Maoris)	Stomach and upper alimentary tract cancers, urinary bladder cancer

Based on the above issues relevant to the safety of herbal medicines, it is clear that the assessment of toxicity at pre-clinical and clinical stages and post-marketing pharmacovigilance evaluation are essential to ensure the safety and effectiveness of herbal medicines [41].

10.4 Regulation of herbal medicine

The massive demand for medicinal plant materials, plant extracts, essential oils, gums, tannins, etc. has resulted in a huge trade at the international level. China and India from Asia; Egypt and Morocco from Africa; Poland, Bulgaria and Albania from Europe; Chile and Peru from South America are some of the important suppliers while the USA, Japan and Europe are the major consumers of those products [53]. The international trade of herbal medicinal products must be conducted in compliance with the Convention on International Trade in Endangered Species of Wild Fauna and Flora (CITES) that regulates the international trade of species that are threatened with extinction or which would be reaching the status of being endangered [54].

The growing trend of consumer acceptance and industrial interests has created a good market for herbal medicinal products. Because of the extensive utility of these products, a more stringent regulatory framework is required to ensure quality and safety. The regulation of herbal medicines is a complex and constantly evolving process that varies from country to country. A particular herbal medicinal product marketed as a "drug" in one country could be used as a "dietary supplement" in another country. An example is *Ginkgo biloba* which was considered as a food in the UK until 2008 while being regulated as a medical product in Germany and as a food supplement in the USA. Nowadays it is regulated as a traditional herbal medicinal product in the UK and several other European countries [55]. As herbal products are classified under different categories (e.g., complimentary medicines, natural health products, prescription medicines, over the counter medicines, supplements, traditional herbal medicines), the regulatory requirements also vary depending on the category to which a particular product belongs. For example, prescription medicines are subjected to strict regulation while the extent of control on supplements is rather low [54]. The strong connection between traditional herbal medicines and indigenous knowledge may result in a lower level of regulations particularly when those are used for the treatment of minor disorders.

The manufacturing process is a major stage where quality control is required. Good manufacturing practices (GMP) are the most important tools to warrant the quality of herbal medicine. GMP for herbal medicine cover all facets from cultivation in the field to the preparation of different formulations [56]. Imposing regulatory standards on herbal medicine to be manufactured using GMP would be helpful to ascertain the general public's belief that herbal medicines are safer than synthetic medicines [57].

10.5 Conclusion

Plant based-medicines and other botanicals play an imperative role in the healthcare systems of many countries. However, the presence of adulterants, contaminants, or inherent toxicity of plant ingredients can lead to adverse effects varying from minor to severe and sometimes fatal. Therefore strict regulations are required to ensure the quality, safety, and efficacy of these herbal medicinal products.

References

[1] World Health Organization. National policy on traditional medicine and regulation of herbal medicines: Report of a WHO global survey. 2005 https://apps.who.int/iris/handle/10665/43229

[2] Phua DH, Zosel A, Heard K. Dietary supplements and herbal medicine toxicities-when to anticipate them and how to manage them. Int J Emerg Med 2009, 2(2), 69–76.

[3] Mosihuzzaman M, Choudhary MI. Protocols on safety, efficacy, standardization, and documentation of herbal medicine (IUPAC Technical Report). Pure Appl Chem 2008, 80(10), 2195–2230.

[4] Choudhary N, Sekhon B. An overview of advances in the standardization of herbal drugs. J Pharm Educ Res 2011, 2(2), 55–70.

[5] Posadzki P, Watson L, Ernst E. Contamination and adulteration of herbal medicinal products (HMPs): An overview of systematic reviews. Eur J Clin Pharmacol 2013, 69(3), 295–307.

[6] World Health Organization, WHO guidelines for assessing quality of herbal medicines with reference to contaminants and residues. Geneva, Switzerland, WHO Press, 2007.

[7] Noor R, Huda N, Rahman F, Bashar T, Munshi SK. Microbial contamination in herbal medicines available in Bangladesh. Bangladesh Med Res Counc Bull 2013, 39(3), 124–129.

[8]   de Sousa Lima CM, Fujishima MAT, de Paula Lima B, Mastroianni PC, Ffo DS, Da Silva JO. Microbial contamination in herbal medicines: A serious health hazard to elderly consumers. BMC Complement Med Ther 2020, 20(1), 17. doi: 10.1186/s12906-019-2723-1.

[9]   Yesuf A, Wondimeneh Y, Gebrecherkos T, Moges F. Occurrence of potential bacterial pathogens and their antimicrobial susceptibility patterns isolated from herbal medicinal products sold in different markets of Gondar Town, Northwest Ethiopia. Int J Bacteriol 2016, 2016, 1959418. doi: 10.1155/2016/1959418..

[10]   Kalumbi MH, Likongwe MC, Mponda J, Zimba BL, Phiri O, Lipenga T, Mguntha T, Kumphanda J. Bacterial and heavy metal contamination in selected commonly sold herbal medicine in Blantyre, Malawi. Malawi Med J 2020, 32(3), 153–159.

[11]   Ting A, Chow Y, Tan W. Microbial and heavy metal contamination in commonly consumed traditional Chinese herbal medicines. J Tradit Chin Med 2013, 33(1), 119–124.

[12]   Altyn I, Mycotoxin TM. Contamination concerns of herbs and medicinal plants. Toxins (Basel) 2020, 12(3), 182. doi: 10.3390/toxins12030182..

[13]   Singh P, Srivastava B, Kumar A, Dubey NK. Fungal contamination of raw materials of some herbal drugs and recommendation of *Cinnamomum camphora* oil as herbal fungitoxicant. Microb Ecol 2008, 56(3), 555–560.

[14]   Trucksess MW, Scott PM. Mycotoxins in botanicals and dried fruits: A review. Food Addit Contam Part A Chem Anal Control Expo Risk Assess 2008, 25(2), 181–192.

[15]   Yuan X, Chapman RL, Wu Z. Analytical methods for heavy metals in herbal medicines. Phytochem Anal 2011, 22(3), 189–198.

[16]   Behera B, Bhattacharya S. The importance of assessing heavy metals in medicinal herbs: A quantitative study. TANG: Int J Genuine Trad Med 2016, 6(1), 1–4.

[17]   Nkansah MA, Hayford ST, Borquaye LS, Ephraim JH. Heavy metal contents of some medicinal herbs from Kumasi, Ghana. Cogent Environ Sci 2016, 2(1), doi: 10.1080/23311843.2016..

[18]   Garvey GJ, Hahn G, Lee RV, Harbison RD. Heavy metal hazards of Asian traditional remedies. Int J Environ Health Res 2001, 11, 63–71.

[19]   Street RA. Heavy metals in medicinal plant products-An African perspective. S Afr J Bot 2012, 82, 67–74.

[20]   Kowalska G. Pesticide residues in some Polish herbs. Agriculture 2020, 10(5), 154. doi: 10.3390/agriculture10050154.

[21]   Tripathy V, Saha A, Kumar J. Detection of pesticides in popular medicinal herbs: A modified QuEChERS and gas chromatography-mass spectrometry based approach. J Food Sci Technol 2017, 54(2), 458–468.

[22]   Tripathy V, Saha A, Patel DJ, Basak BB, Shah PG, Kumar J. Validation of a QuEChERS-based gas chromatographic method for analysis of pesticide residues in *Cassia angustifolia* (Senna). J Environ Sci Health- Part B 2016, 51, 508–518.

[23]   Opuni KFM, Asare-Nkansah S, Osei-Fosu P, Akonnor A, Bekoe SO, Dodoo ANO. Monitoring and risk assessment of pesticide residues in selected herbal medicinal products in Ghana. Environ Monit Assess 2021, 193, 470. doi:10.1007/s10661-021-09261-1.

[24]   Xu M, Huang B, Gao F, Zhai C, Yang Y, Li L, Wang W, Shi L. Assessment of adulterated traditional Chinese medicines in China: 2003–2017. Front Pharmacol 2019, 10, 1446. doi: 10.3389/fphar.2019.01446.

[25]   Yu C, Wang CZ, Zhou CJ, Wang B, Han L, Zhang CF, Wu XH, Yuan CS. Adulteration and cultivation region identification of American ginseng using HPLC coupled with multivariate analysis. J Pharm Biomed Anal 2014, 99, 8–15.

[26]   Cui S, Wu J, Wang J, Wang X. Discrimination of American ginseng and Asian ginseng using electronic nose and gas chromatography-mass spectrometry coupled with chemometrics. J Ginseng Res 2017, 41(1), 85–95.

[27] Lee J, Shibamoto T, Ha J, Jang HW. Identification of volatile markers for the detection of adulterants in red ginseng (*Panax ginseng*) juice using headspace stir-bar sorptive extraction coupled with gas chromatography and mass spectrometry. J Sep Sci 2018, 41, 2903–2912.

[28] Hwang SG, Kim JH, Moon JC, Kim J-H, Jang CS. Comparative analysis of chloroplast DNA sequences of *Codonopsis lanceolata* and *Platycodon grandiflorus* and application in development of molecular markers. Appl Biol Chem 2017, 60, 23–31.

[29] Heubl G. DNA-based authentication of TCM-plants: Current progress and future perspectives. In: Wagner H, Ulrich-Merzenich G, eds, Evidence and rational based research on Chinese drugs. 1st ed, Vienna, Austria, Springer-Verlag, 2013, 27–85.

[30] Koh HL, Wang H, Zhou S, Chan E, Woo SO. Detection of aristolochic acid I, tetrandrine and fangchinoline in medicinal plants by high performance liquid chromatography and liquid chromatography/mass spectrometry. J Pharm Biomed Anal 2006, 40(3), 653–661.

[31] But PP. Herbal poisoning caused by adulterants or erroneous substitutes. J Trop Med Hyg 1994, 97(6), 371–374.

[32] But PP, Tomlinson B, Cheung KO, Yong SP, Szeto ML, Lee CK. Adulterants of herbal products can cause poisoning. BMJ. 1996, 313(7049), 117. doi: 10.1136/bmj.313.7049.117a..

[33] Zhao Z, Hu Y, Liang Z, Yuen JP, Jiang Z, Leung KS. Authentication is fundamental for standardization of Chinese medicines. Planta Med 2006, 72(10), 865–874.

[34] Zhu YP. Toxicity of the Chinese herb mu tong (*Aristolochia manshuriensis*); What history tells us. Adverse Drug React Toxicol Rev 2002, 21(4), 171–177.

[35] Huang WF, Wen KC, Hsiao ML. Adulteration by synthetic therapeutic substances of traditional Chinese medicines in Taiwan. J Clin Pharmacol 1997, 37, 344–350.

[36] Ko RJ. Causes, epidemiology, and clinical evaluation of suspected herbal poisoning. J Toxicol Clin Toxicol 1999, 37, 697–708.

[37] Firozian F, Amir NA, Moradkhani S, Moulaei M, Fasihi Z, Ahmadimoghaddam D. Adulteration of the herbal weight loss products by the illegal addition of synthetic antiobesity medications: A pilot study. J Obes 2021, 2021, 9968730. doi:10.1155/2021/9968730.

[38] Haj AL, Al Busaidi I, Kadavilpparampu AM, Suliman FEO. Determination of common adulterants in herbal medicine and food samples using core-shell column coupled to tandem mass spectrometry. J Chromatogr Sci 2017, 55(3), 232–242.

[39] Primpray V, Chailapakul O, Tokeshi M, Rojanarata T, Laiwattanapaisal WA. Paper-based analytical device coupled with electrochemical detection for the determination of dexamethasone and prednisolone in adulterated traditional medicines. Anal Chim Acta 2019, 1078, 16–23.

[40] Ernst E. Adulteration of Chinese herbal medicines with synthetic drugs: A systematic review. J Intern Med 2002, 252(2), 107–113.

[41] Moreira DDL, Teixeira SS, Monteiro MHD, Acax D-O, Paumgartten FJR. Traditional use and safety of herbal medicines-1. Rev Bras Farmacogn 2014, 24(2), 248–257.

[42] Takeuchi S, Homma M, Inoue J, Kato H, Murata K, Ogasawara T. Case of intractable ventricula fibrillation by a multicomponent dietary supplement containing ephedra and caffeine overdose. Chudoku Kenkyu 2007, 20(3), 269–271.

[43] Boumba VA, Mitselou A, Vougiouklakis T. Fatal poisoning from ingestion of *Datura stramonium* seeds. Vet Hum Toxicol 2004, 46(2), 81–82.

[44] Liu Q, Zhuo L, Liu L, Zhu S, Sunnassee A, Liang M, Zhou L, Liu Y. Seven cases of fatal aconite poisoning: Forensic experience in China. Forensic Sci Int 2011, 10 212(1–3), e5–9.

[45] Ernst E. Toxic heavy metals and undeclared drugs in Asian herbal medicines. Trends Pharmacol Sci 2002, 23, 136–139.

[46] Boullata J. Natural health product interactions with medication. Nutr Clin Pract 2005, 20(1), 33–51.

[47] Chua YT, Ang XL, Zhong XM, Khoo KS. Interaction between warfarin and Chinese herbal medicines. Singapore Med J 2015, 56(1), 11–18.

[48] Haller CA. Clinical approach to adverse events and interactions related to herbal and dietary supplements. Clin Toxicol 2006, 44(5), 605–610.

[49] Borrelli F, Izzo AA. Herb-drug interactions with St John's wort (*Hypericum perforatum*): An update on clinical observations. AAPS J 2009, 11(4), 710–727.

[50] Poivre M, Nachtergael A, Bunel V, Philippe ON, Duez P. Genotoxicity and carcinogenicity of herbal products. In: Pelkonen O, Duez P, Vuorela P, Vuorela H, eds, Toxicology of herbal products. 1st ed, Cham, Switzerland, Springer, 2017, 179–215.

[51] Popat A, Shear NH, Malkiewicz I, Thomson S, Neuman MG. Mechanism of impila (*Callilepis laureola*)-induced cytotoxicity in Hep G2 cells. Clin Biochem 2002, 35, 57–64.

[52] Sheikh NM, Philen RM, Love LA. Chaparral-associated hepatotoxicity. Arch Intern Med 1997, 157(8), 913–919.

[53] Heinrich M. Quality and safety of herbal medical products: Regulation and the need for quality assurance along the value chains. Br J Clin Pharmacol 2015, 80(1), 62–66.

[54] Sharma S. Current status of herbal product: Regulatory overview. J Pharm Bioallied Sci 2015, 7(4), 293–296.

[55] Vasisht K, Sharma N, Karan M. Current perspective in the international trade of medicinal plants material: An update. Curr Pharm Des 2016, 22(27), 4288–4336.

[56] Mukherjee PK. Problems and prospects for good manufacturing practice for herbal drugs in Indian systems of medicine. Drug Inf J 2002, 36(3), 635–644.

[57] Chan K. Some aspects of toxic contaminants in herbal medicines. Chemosphere 2003, 52(9), 1361–1371.

Jagadeshwar Reddy Thota, Ravi Kumar Maddula,
Sukanya Pandeti, Naga Veera Yerra

11 Advances in extraction and analysis of natural products

11.1 Introduction

There is enormous potential in traditional medicine systems for curing a variety of ailments and diseases in humans [1]. Plants and plant products have been a treasured source for novel molecules contemplated as substitute scenarios in the discovery of innovative drugs. Usage of natural products possessing therapeutic values has been potentially applied in curing diseases and constituted in ancient traditional civilization. However, extraction, separation, isolation, and analytical sequestration of natural products still remain a herculean task. A predominant quantity of natural products from plants/herbs is extracted from leaves, seeds, flowers, bark, and roots [2, 3]. An advantage of extracting natural products from plant sources is their inconceivable bioactive characteristics like insecticides, fungicides, antioxidants, and growth promoters. Furthermore, the increase in consumption of natural products will help in containing the detrimental impact of synthetic products and further curb their usage [4].

A variety of applications such as phytochemicals, cosmetics, food processing, lipids, pharmaceuticals, flavors, fragrances, pesticides, and pigments have been recorded for the valuation of these natural products. Being a renewable resource along with biodegradable nature, natural products make a superior quality and economical alternative with sustainable growth. The varying distribution of secondary metabolites in plants and associated biological functions are definitive to that plant in which they are detected [5]. These metabolites are habitually accountable for characteristics like flavor, color, taste, and fragrance which typically resolve the plant-environment interaction. Some metabolites such as polyphenols, alkaloids, terpenes, polyketides, and hormones are associated with the defense and signaling mechanisms of the plant which results in the diversity of its chemical and molecular structure existing in the plant.

Conventional extraction methods like maceration and Soxhlet have been in use for decades which are tedious, time-consuming, and employ a reasonably large quantity of organic solvents [6]. This makes clear that the new extraction systems with decreased extraction time and solvent, rugged, and high-yielding extraction methods are crucial. Of late new extraction techniques such as ultrasound-assisted extraction,

Jagadeshwar Reddy Thota, CSIR-Indian Institute of Chemical Technology, Tarnaka, Hyderabad 500 007, Telangana, India, e-mail: tjreddy@iict.res.in
Ravi Kumar Maddula, Sukanya Pandeti, Naga Veera Yerra, CSIR-Indian Institute of Chemical Technology, Tarnaka, Hyderabad 500 007, Telangana, India

https://doi.org/10.1515/9783110595949-011

microwave-assisted extraction (MAE), supercritical fluid extraction, and accelerated solvent extraction are dominating for competent and quick extraction from complex plant matrices. The advantage of these techniques to work at elevated temperatures and/or pressures relatively decreases the time of extraction [7].

Natural product characterization in analytical chemistry using prevailing techniques such as mass spectrometry (MS) and nuclear magnetic resonance (NMR) spectroscopy (NMR) are represented as structure elucidation tools without prior isolation of natural products [8]. Nevertheless, in many samples, it becomes inevitable to purify the compound before analysis. To properly understand the chemical structure and stereochemistry, the new natural products are required to be isolated and available as high purity compounds. Still, this is not a shortcoming because the sample amounts needed are relatively less and techniques like 2D NMR spectra can be obtained with just 100 mg within feasible time. In addition to this, the isolated natural product has to be subjected to biological activity studies *in vitro* and *in vivo* after purification to eliminate the interference of accompanying compounds [9]. Those certified reference standards used for testing quality control of herbal extracts principally rely on isolated compounds with recorded purity. Natural products in recent times have proficient rejuvenation in drug-discovery programs due to their superior natural diversity over synthetic compounds and their structural lookalike. The isolation of natural products begins with the collection, identification, and preparation of biological material typically by heat treatment/drying. This follows extraction using different solvents from low to high polarity. Before proceeding to the isolation of pure compounds, (semi-)preparative HPLC or liquid–liquid chromatographic techniques and defined purification steps are essential to eradicate the unwanted matrix [10].

11.2 Sample preparation methods

Sample extraction is one of the pivotal and crucial steps which is a comprehensive task in natural products chemistry. The selection of the extraction process is subjected to the nature of the source material from which compounds are to be isolated. Extraction of desired organic components from the plant material and their separation and structural characterization is mandatory before proceeding to choose a method. Sample pre-concentration benefits to enhance the efficiency of analysis, reduce potential interferences, and augment sensitivity, reliability, accuracy, and reproducibility of the analysis by amplifying the analyte concentration in the assay mixture. Further, sample preparation also transforms desired analytes into more relevant ones that can be easily separated, detected, and quantified. Consequently, the obtained sample should possess high concentrations of target analytes free from any background matrices; hence, the vital step in this process is the extraction of a target analyte [7, 11, 12]. Traditional extraction procedures applied to plant materials extraction like maceration or Soxhlet typically use a huge amount of organic solvents with long extraction periods. There are certain limitations of these traditional extractions like drying of the extract after solvent evaporation, chemical transformation during the high-temperature and/or lengthy extraction, along with the generation of large volumes of toxic waste. However, many state-of-art extraction techniques available enjoy certain advantages like reduced organic solvent consumption, improved extraction efficiency, and selectivity [4].

11.2.1 Pros and cons of innovative extraction methods

The choice of extraction method to be applied for a particular matrix relies on the nature of raw material to be processed and the product desired. There cannot be a uniform and definitive extraction method for obtaining desired bioactive compounds from natural products; each method carries its own limitations and advantages [7]. Predominantly compound isolation processes still employ procedures using organic solvents of diverse polarity, water, and their mixtures. Conventional extraction techniques for bioactive compounds are stationed on the choice of solvent and the use of heat and/or agitation [13]. The most pronounced conventional techniques are maceration, soaking extraction, Soxhlet extraction, and distillation [14]. These methods are still in practice because of their painless and economical ways to obtain essential oils and bioactive compounds from plant material. Nevertheless, the extraction durations are relatively very long: Soxhlet extraction typically lasts from 4 to 48 hrs. Other liquid extraction techniques like wrist shaker or hot-plate boiling require large solvent amounts and are often time-consuming (filtering, pre-concentration before analysis) [4].

11.2.1.1 Maceration

This is a simple, widely used extraction procedure for thermolabile compounds, e.g., terpenoids, phenolics, alkaloids, [15, 16]. This method is appropriate for both simple and bulk extractions. The procedure involves pulverized plant material to soak in a suitable solvent in a closed container at room temperature with eventual stirring. The extraction comes to a halt when equilibrium is attained between the concentration of metabolites in the extract and that of plant material. However, the main limitation is this extraction procedure consumes large volumes of solvent along with long extraction times [15–20].

11.2.1.2 Soxhlet extraction

This is one of the widely practiced methods for decades for the extraction of nutraceuticals from diverse plant matrices [15]. The Soxhlet extraction applied for bioactive compounds harmonizes the advantage of reflux extraction and percolation using the principle of reflux and siphoning to constantly extract the plant material with fresh solvent. In the Soxhlet extraction, the selected nutraceuticals are extracted from finely ground plant material by using suitable solvents or a mixture of solvents under heating reflux conditions. Hexane can be applied in the foremost step to eliminate chlorophylls and fatty components. The main advantage of Soxhlet extraction is that it is a continuous process with high extraction efficacy that desires less time and solvent consumption than maceration or percolation. Even though the extraction takes several hours, the yields are higher than contemporary methods like microwaves or ultrasound-assisted extraction. However, the main disadvantage of Soxhlet extraction is that the extract heated constantly at the boiling point of the solvent can damage thermolabile compounds and/or initiate the forming of artifacts [4, 21, 22].

11.2.1.3 Percolation

In the percolation extraction, nutraceuticals are extracted from powdered plant material by soaking in a suitable solvent in a percolator. In general, the extraction efficiency is more in percolation than maceration as soaking solvent is replaced with fresh solvent continuously during the extraction process. Optimization of the percolation method(s) by way of using different percentages of extraction solvents and extraction time are the crucial steps in this process [23, 24]. The major limitation of this method is extraction at higher temperatures may lead to loss or decomposition of labile metabolites, and this method consumes a large volume of solvents and also time.

11.2.1.4 Reflux extraction

Reflux is one of the methods for the extraction of thermally stable natural products. In this method, plant material is refluxed with a suitable solvent for the extraction of nutraceuticals. The extraction efficiency depends on the solvent used, temperature, and time. It requires less solvent and extraction time than maceration or percolation methods [7, 25].

11.2.1.5 Microwave-assisted extraction

Microwave-assisted extraction (MAE) is one the classical extraction technique that is used for the extraction of medicinally active components from various plant matrices. In this technique, the microwave energy is used to heat the sample solvent mixture rapidly resulting in the partitioning analytes from a sample matrix in to the solvent. The extraction involving microwave radiation by interacting with polar compounds such as water along with organic components in the plant matrix following the ionic conduction and dipole rotation mechanism augments the recovery of secondary metabolites and aroma compounds. This process is usually performed using microwave energy ranges of 300 MHz to 300 GHz [26]. The ease and convenience of carrying out MAE under an inert atmosphere decrease the probabilities of degradation of sensitive compounds due to high temperature and high irradiation [15, 27]. It is also considered to be a green extraction technique due to shortened extraction time and solvent consumption.

11.2.1.6 Ultra sonication–assisted extraction

Ultrasonic-assisted extraction (UAE) or sonication is a technique in which ultrasound wave energy is used in extraction typically in the frequency ranging between 20 kHz and 100 MHz. Similar to other waves, it passes through a medium by creating compression and expansion producing cavitation, meaning growth and collapse of bubbles. Cavitation fast-tracks the diffusion and dissolution of the solute along with heat transfer which improves the extraction efficiency. As per Doktycz and Suslick [28], obtained bubbles have a temperature of about 5,000 K, pressure 1,000 atm, and heating and cooling rate above 1,010 K/s. The chief benefit of the UAE is that ultrasound wave energy facilitates both organic and inorganic compounds leaching from the plant matrix. The escalation of mass transfer and accelerated access of solvent to cell materials of plant parts is the apparent scheme behind UAE [7]. UAE involves two specific physical phenomena: (a) the diffusion across the cell wall and (b) rinsing the contents of the cell after breaking the walls [4, 15, 29]. The low solvent and energy consumption, and the reduction of extraction temperature and time, make UAE advantageous over other extraction techniques. Further, the applicability of UAE for thermolabile and unstable compounds makes it preferential. UAE is widely applied to facilitate the extraction of intracellular metabolites from plant cell cultures [30].

11.2.1.7 Pressurized liquid extraction

Pressurized liquid extraction (PLE), also popular as accelerated solvent extraction, applies high pressure in extraction. This method is quite similar to supercritical fluid extraction, wherein high pressure keeps solvent in a liquid state above their boiling points which results in high solubility and diffusion rate of solute into the solvent with high penetration of solvent into the matrix. PLE requires specialized, expensive equipment for maintaining high pressure and temperature conditions to increase efficiency, selectivity, and repeatability. The reduced extraction time and solvent consumption over other methods make PLE an economical and environmentally friendly alternative to conventional extraction techniques. Conversely, the PLE equipment is less user-friendly in terms of sample preparation time and workforce [4, 7].

11.2.1.8 Accelerated solvent extraction

Accelerated solvent extraction (ASE) is a fully advanced closed rapid extraction technique for the extraction of organic compounds from solid and semi-solid matrices. The procedure briefly comprises three successive steps, i.e., loading of the sample into the extraction cell, filling the solvent into the cell, and finally purging the residual extract. In situ derivatization is also possible during the extraction process. ASE activates at a temperature above the normal boiling point of solvents and uses pressure to keep solvent in liquid form during extraction. The increase in temperature accelerates the extraction capacity of the solvent to solubilize the analytes and elevated pressure pumps the solvent through the matrix bed resulting in more close contact with the analytes. The elevated pressure also keeps the solvent below its boiling point enabling rapid, safe, and efficient extraction of target compounds from various matrices. The extraction can be effectively performed at selected operational temperature and pressure [15, 31]. Once the extraction is completed, the residual extract is purged using an inert gas such as nitrogen. The time required for ASE is short compared to that of conventional maceration and Soxhlet techniques [4]. This technique can be conveniently scaled up and applied in natural products, food materials, and agricultural residues. The noticeable leverage of ASE includes extraction for sample sizes 1–100 g, relative solvent reduction, wide application range, and managing acidic and alkaline matrices. It is also supposed to offer lower-cost per sample than other extraction techniques [32].

11.2.1.9 Supercritical fluid extraction

Supercritical fluid extraction (SFE) uses supercritical fluid as the extraction solvent and is an eco-friendly technique [33]. Supercritical fluid has similar solubility to liquid and similar diffusivity to gas which can dissolve a wide range of natural products. Supercritical carbon dioxide is widely used in SFE due to highly remarkable advantages like low critical temperature (31 °C), selectivity, inertness, low cost, non-toxic, and specific capability in extracting thermolabile compounds. Further, the addition of a modifier to supercritical carbon dioxide will enhance the solvating properties significantly. The prime superiority of SFE includes: (1) the dissolving power of a supercritical fluid solvent is changeable by altering pressure and temperature. (2) the supercritical fluid has a higher diffusion coefficient and lower viscosity and surface tension than a liquid solvent, leading to more favorable mass transfer [15, 29]. This method has proven to be more advantageous due to the process conducted at low temperature which prevents the damage of sensitive compounds along with amounts of solvent residues. SFE is an impressive alternative with reduced consumption of organic solvent and an eco-friendly process that can be applied for extracting pesticides, dietary supplements, fragrances, and natural products [7, 34].

11.2.1.10 Hydrodistillation

The hydrodistillation method is commonly used for the extraction of volatile oil from plant materials [35]. This method is subcategorized into water distillation, steam distillation, or a combination of water and steam distillation. In this method, volatile compounds; i.e., primary and secondary essential oils are vaporized along with the steam of water and then condensed at low temperatures which results in an immiscible mixture of an oil phase and an aqueous phase. The major drawback of this method is the long periods of extraction and hydrolytic or thermal decomposition of ester or unsaturated natural compounds [36].

11.2.1.11 Enzyme-assisted extraction

The enzyme-assisted extraction method is specific and efficient [7, 37]. Different types of enzymes are used to treat plant material to extract specific bioactive natural products from plant materials. In this method, plant cell walls are broken down and the metabolites are released. Generally, these enzymes are derived from different sources, e.g. vegetables, animal organs, bacteria, fruits, and fungi. Based on the catalytic activity of the enzymes, the enzymes are categorized into different types which are ligases, oxidation-reduction enzymes, group transfer enzymes, desmolases, and carboxylation enzymes, hydrolyzing enzymes, isomerizing enzymes, etc. This method requires less time and organic solvents to extract natural products with high yield and purity [38, 39]. The major limitations of this method are mainly: (1) enzymes are expensive to use on an industrial scale, (2) presently available enzymes cannot rupture the plant cell walls completely, (3) the behavior of enzymes varies with different environmental conditions.

A brief summary of various extraction methods for natural products is given in Table 11.1.

Table 11.1: Various extraction methods for natural products.

S. no	Extraction method	Extraction solvent	Temperature/ pressure	Extraction time	Expected natural product
1	Maceration	Water, aqueous and non-aqueous solvents	RT/AP	Long	Based on extraction solvent polarity
2	Soxhlet extraction	Organic solvents	UH/AP	Long	Based on extraction solvent polarity
3	Percolation	Water, aqueous and non-aqueous solvents	RT or UH/AP	Long	Based on extraction solvent polarity
4	Reflux extraction	Aqueous and non-aqueous solvents	UH/AP	Moderate	Based on extraction solvent polarity
5	Microwave- assisted extraction	Water, aqueous and non-aqueous solvents	UH/AP	Short	Based on extraction solvent polarity
6	Ultra sonication– assisted extraction	Water, aqueous and non-aqueous solvents	RT or UH/AP	Short	Based on extraction solvent polarity
7	Pressurized liquid extraction	Water, aqueous and non-aqueous solvents	UH/HP	Short	Based on extraction solvent polarity
8	Accelerated solvent extraction	Organic solvents	Elevated temperature and pressure	Short	Based on extraction solvent polarity
9	Supercritical fluid extraction	Supercritical fluid (usually $S\text{-}CO_2$), sometimes with modifier	RT/HP	Short	Nonpolar to moderate polar compounds
10	Hydro distillation	Water	UH/AP	Long	Essential oil
11	Enzyme- assisted extraction	Water, aqueous and non-aqueous solvents	RT/AP	Moderate	Based on extraction solvent polarity

RT-Room Temperature; UH-Under Heat; AP-Atmospheric Pressure; HP-High Pressure

11.3 Analytical techniques for natural products

Essential parameters for an analytical technique for standardizing medicinal plants and plant products include high sensitivity, improved resolution, clean separation, and lowest detection limits added with minimum analysis time. Application of the right analytical technique has become decisive for quantifying environmental, pharmaceutical, toxicological, natural products, polymers, and chemical synthesis samples. To cater to these needs, techniques that syndicate both chromatographic methods (LC, GC, CE, etc.) with diverse spectroscopic methods (UV, MS, NMR, etc.) have emerged [10]. This new outcome of coupling (hyphenation) has given a new dimension in the field of separation and purification analysis which is efficient, precise, adjustable, and can be oriented to any specific analytical application.

Equitable and prudent analytical techniques contribute a vital segment in the discovery of novel active compounds of natural products along with solving intricate analytical problems.

11.3.1 Thin-layer chromatography

Thin-layer chromatography (TLC) is one of the widely used methods for the separation and primary identification of constituents of plant extracts. In TLC, compound separation is accomplished by partition and adsorption depending on the composition of the stationary phase and the mobile phase. The cost-effective nature of the equipment and consumables makes it versatile in application. Natural product analysis using TLC mainly centers on material identification, monitoring the progress of a reaction and detection of tainted compounds. As most of the compounds appear under visible region, TLC becomes an easy technique for detecting bioactive compounds; furthermore applying different derivatization reactions helps to improve the effectiveness of purification and determination [40, 41].

11.3.2 High-performance thin-layer chromatography

High-performance thin-layer chromatography (HPTLC) is a sophisticated planar chromatography and most advanced system of instrumental TLC. HPTLC gives better separation, increased resolution, more accurate quantitative measurements through low analysis time [42], less cost per analysis with little maintenance cost. Moreover, the hyphenation of HPTLC with mass spectrometry (HPTLC-MS), Fourier transform infrared spectroscopy (HPTLC-FTIR), laser desorption, scanning diode, etc. has shown its wider capability in analytical determinations. Simultaneous processing of standard and sample with improved analytical accuracy and precision are making HPTLC and its hyphenated techniques a substantial tool in fields like pharmaceuticals analysis, natural products chemistry, and pharmacokinetics [43].

11.3.3 Counter-current chromatography

"Counter-current chromatography" (CCC) is an initiatory generic term covering all forms of liquid–liquid chromatography that uses support free liquid stationary phase. CCC uses an immiscible biphasic liquid system with one liquid being a stationary phase and the other being a mobile phase [44]. The working principle is based on the countercurrent partition system, wherein the partition of solute particles happens between two immiscible solvents. This separation is based on the different "partitioning coefficients" of a particular compound present in the solvent phase versus the diluent phase. This is similar to the HPLC technique except for holding a liquid stationary is more difficult than solid. The stationary phase which is let into the column whirls at reasonable rotating speed and is held by generated centrifugal force. The mobile phase containing solute particles intend to be separated is fed into the column and then pushed to the stationary phase. Nevertheless, the separation efficacy can be less due to poor mixing of two phases that improves the mixing of solvents introduced high-speed CCC technique [4, 45].

The advantage of CCC lies in its high resolution and separation capability resulting in high sample recovery. Based on the sample solubility and the column volume, CCC can be adapted to large sample volumes [10]. Droplet-CCC (DCCC) has found wide applications in the preparative separation of plant constituents and other natural products with particular reference to the isolation of polar compounds [46]. Effective application of the CCC technique includes the separation of essential oils, steroids, plant growth regulators, alkaloids, glycosides, and antibiotics.

11.3.4 High-performance liquid chromatography

High-performance liquid chromatography (HPLC) has been established as a paramount and highly adapted chromatographic technique for the separation of natural products in crude and complex matrices. HPLC has become pioneering analytical support for the identification, quantification, and purification of individual components from a mixture [47]. An HPLC system characteristically comprises following components: (i) solvent reservoir; (ii) pump system; (iii) sample injection system; (iv) column compartment; and (v) detector. Instead of the mobile phase being allowed to drip through the column under gravity, it is forced under high pressure up to 400 atmospheres to make the separation faster. Ultra-high-pressure liquid chromatography enhances mainly the speed, resolution, and sensitivity, and is capable of coping with high backpressures which resulted in remarkable improvements in the analysis of complex plant extract mixtures. The diversity of natural products makes it difficult in selecting one particular detection system. Moreover, there is no single detection technique for their effective determination. The most widely used HPLC detectors include ultraviolet-visible (UV), diode-array detectors (DAD), fluorimetric detectors (FLD), electrochemical detectors (ECD), refractive index detectors (RID), chemiluminescence (CL) detectors (CL), evaporative light scattering detectors (ESLD), and charged aerosol detector (CAD) [48].

11.3.4.1 Liquid chromatography with ultraviolet-visible detection (HPLC-UV)

In recent times LC-UV, with single-wavelength or diode array detectors (DAD), is becoming the most popular technique for the separation of compounds due to the relatively economical and effortless operation of the instrument. Despite some limitations, like for natural products that do not possess UV chromophores, it has the best combination of sensitivity, linearity, versatility, and reliability of all available HPLC detectors. Most natural products absorb UV in the range of 200–550 nm, all constituents having one or more double bonds and ingredients having unshared electrons. Therefore, compounds having weak chromophores can be successfully detected by UV at short wavelengths. The UV detector is generally set at 254 nm, but it can be used at more specific wavelengths such as 280 nm, 430 nm, 480 nm, or 500 nm [49, 50].

11.3.4.2 Liquid chromatography with fluorimetric detection (HPLC-FLD)

Fluorimetry detection (FLD) is not widely used, but this technique is known for its high sensitivity for a selective group of compounds. By using a specific wavelength, analyte atoms are excited and then emit a light signal; the intensity of emitted light is monitored to quantify the analyte of interest. Hence the natural products can be detected without derivatization. It has been documented that HPLC-FLD is suitable for routine quality assurance to control the presence of natural compounds from the raw material [4].

11.3.4.3 Liquid chromatography coupled to mass spectrometry (HPLC-MS)

The separation capabilities of LC and the identification power of MS combination evolved as a predominant tool in the analysis of plant extracts. The MS detector proficiency to separate gas phase ions according to their *m/z* (mass to charge ratio) value makes it unique and beneficial. The possibility of chemical structure elucidation of compounds using tandem analysis makes it exceptional. Using tandem MS, some of these ions are selected and new fragments ions are produced which pave the way in determining the chemical structure of analytes. Moreover, MS detector improves the signal/noise ratio and selectivity by providing specific modes such as single ion monitoring and selected reaction monitoring, and by separating co-eluted compounds [4, 51].

Furthermore, electrospray ionization (ESI) became a widely accepted choice as this ionization source is suitable for a large range of molecules, making it an utmost acceptable technique to the diversity of polarities and weights encountered among secondary metabolites [10].

Some auxiliary detectors with different detection modes attached to liquid chromatography have also been employed in the analysis of biological matrices (other than plant materials). Albeit HPLC-UV/HPLC-PDA are widely applied techniques due to their simple handling, lack of sensitivity, selectivity, and interference of the co-eluting matrix still remain a challenge (e.g., detection of anthraquinone) [4]. Mass spectrometry detection is undoubtedly an analytical tool for facile and speedy determination and quantification. Meanwhile, the significant expenditure is still limiting its use; therefore, a reasonably priced mass spectrometry can become a strong podium for structure elucidation workflows in the field of natural products research.

11.3.4.4 HPLC coupled to an electrochemical detector

HPLC coupled with an electrochemical detector (ECD) is an extremely sensitive and selective detection technique applied in the determination of compounds possessing electrochemical activity. In a reductive mode, degassing instruments are essential to remove dissolved oxygen from the mobile phase to prevent the oxidation of analytes. Classification and modes of ECD comprise amperometric, coulometric, conductimetric, and potentiometric detectors largely used in biomedical analysis [4].

Among other detectors used, CL detection was successfully applied for the analysis of natural products. CL is defined as the production of electromagnetic radiation (ultraviolet, visible, or infrared) observed through an electronically excited intermediate or product due to a chemical reaction. HPLC-CL proved to be a quite beneficial tool due to its simplicity, low cost, and high selectivity and sensitivity [52].

11.3.5 Mass spectrometry

During the last two decades, mass spectrometry (MS) coupled with molecular separation techniques such as gas or liquid chromatography or capillary electrophoresis is being used for the identification of nutraceuticals in complex mixtures of natural product extracts [51]. Day-by-day developments in mass spectrometric and chromatographic techniques have made it possible for qualitative and quantitative analysis of chemical components in different sample matrices at the sub-ppm level. Mass spectrometers having high resolving power mass analyzers such as Orbitrap, Fourier transform ion cyclotron resonance, and time-of-flight provide high-resolution mass data and help in the structural elucidation of non-targeted/unknown nutraceuticals. In addition to the above mass spectrometry techniques, matrix-assisted laser desorption/ ionization mass spectrometry imaging (MALDI-MSI) is being used in plant sciences to assist in situ detection of a variety of molecules on the surface of a tissue section including primary and secondary plant metabolites as well as peptides and proteins [53].

11.3.6 Nuclear magnetic resonance spectroscopy

NMR spectroscopy is being widely used for qualitative and quantitative analysis of natural products [54]. NMR is a versatile tool to characterize natural products that are present in polar, semi-polar, non-polar extracts of plant sample matrices. Polymers such as polysaccharides, and lignin present in plant cell walls can be profiled by using the NMR technique without sample extraction [55]. Structure elucidation of targeted and non-targeted, primary or secondary plant metabolites, natural compounds can be achieved by one-dimensional and multi-dimensional NMR experiments such as ^{1}H NMR, ^{13}C NMR, correlation spectroscopy (COSY), hetero-nuclear multiple-bond correlation (HMBC), hetero-nuclear single quantum correlation (HSQC), etc. Structure conformations and analyzing molecular interactions of the compounds can also be determined by this method [56].

11.3.7 Capillary electrophoresis

Capillary electrophoresis (CE) has proved to be one of the most dynamic analytical techniques for the qualitative and quantitative determination of various classes of natural products with a wide range of polarity. Due to its versatility and high separation efficiency, CE has emerged as an interesting alternative to routine techniques. CE technique is based on the electrophoretic mobility of charged molecules in a conductive medium under an applied voltage [57].

11.3.8 Gas chromatography

Gas chromatography (GC) is a versatile tool in the analysis of volatile plant secondary metabolites. Headspace, solid-phase microextraction and steam distillation extraction can be used to collect the volatile compounds from small amounts of plant material. However, chemical pretreatments such as derivatization reactions are required to improve the volatility [58, 59].

A very powerful combination of the high separation capacity of GC with the identification capabilities of mass spectrometry such as gas chromatography-atomic absorption spectrometry (GC-AAS); gas chromatography-atomic emission spectrometry (GC-AES); gas chromatography-mass spectrometry (GC-MS); or gas chromatography-inductively coupled plasma mass spectrometry (GC-ICP-MS) techniques provides quantitative and qualitative information [60].

11.3.9 IR spectroscopy (IR)

IR spectroscopy has been used for qualitative and quantitative analysis of nutraceuticals in a wide range of plants and agricultural products. This technique is adequate to fingerprint the composition of bioactive compounds present in plant material. IR analysis is rapid, cheaper, and environmental friendly [61]. However, the major limitation of this technique is it cannot measure molecules that are present at lower concentrations in different plant matrices.

11.4 Dereplication

Dereplication is a low-duration process used for quick identification of known nutraceuticals present in plant extracts without compound isolation used in structure elucidation [62]. Of late a variety of dereplication methodologies, for example, biological assays, analytical tools, computer and statistical tools, have been actively applied in the field of bioactive natural product discovery [63]. Documentation of targeted secondary plant metabolites in plain form or associated with complex matrices (crude extract) can be efficiently conceded using sophisticated separation and spectroscopic/spectrometric techniques. Prevalent analytical methods practiced for the dereplication of natural products include molecular separation techniques like liquid chromatography with different detectors, namely, photodiode array (PDA) or ultraviolet (UV), mass spectrometry (MS), and NMR spectroscopy [64].

Even though UV or DAD detectors are restricted to the interpretation of peak retention times or spectral fingerprints in the case of UV, despite these reservations both can be engaged in dereplication strategies for identification and quantification using standard reference molecules. Meanwhile, MS and NMR spectroscopic techniques are the most comprehensively applied techniques for the dereplication of the natural products envisaging data comparison with the database libraries [65]. The additional capability of these techniques enables the identification of known and unknown nutraceuticals in plants efficiently using the algorithms (ACD Labs, XCMS, MET–IDEA, etc.) for automatic and statistical interpretation.

11.5 Concluding remarks

The high throughput, efficient, and robust novel approaches employed in the extraction and analysis of natural products signify that phytochemical screening is no longer as cumbrous as it was a few years ago.

References

[1]   Brown SL. Lowered serum cholesterol and low mood. Br Med J (Clin Res Ed) 1996, 313, 637–638.
[2]   Avery GS, Speight TM Avery's drug treatment Principles and Practice of clinical pharmacology and therapeutics. Auckland, NZ, ADIS Press Ltd. 1987.
[3]   Kong JM, Goh NK, Chia LS, Chia TF. Recent advances in traditional plant drugs and orchids. Acta Pharmacol Sin 2003, 24(1), 7–21.
[4]   Duval J, Pecher V, Poujol M, Lesellier E. Research advances for the extraction, analysis and uses of anthraquinones: A review. Ind Crops Prod 2016, 94, 812–833.
[5]   Croteau R, Kutchan TM, Lewis NG. Natural products (Secondary metabolites). In: Buchanan B, Gruissem W, Jones R, eds Biochemistry & molecular biology of plants. 1st, Rockville, USA, American Society of Plant Physiologist, 2000, 1250–1318.
[6]   Handa SS, Khanuja SPS, Longo G, Rakesh DD. Extraction technologies for medicinal and aromatic plants. Trieste, IT, United Nations Industrial Development Organization and the International Centre for Science and High Technology, 2008.
[7]   Zhang QW, Lin LG, Ye WC. Techniques for extraction and isolation of natural products: A comprehensive review, Chin Med 2018, 13, 20. 10.1186/s13020-018-0177-x.
[8]   Emwas AH, Roy R, McKay RT, Tenori L, Saccenti E, Gowda GAN, Raftery D, Alahmari F, Jaremko L, Jaremko M, Wishart DS. NMR spectroscopy for metabolomics research. Metabolites 2019, 9(7), 123. 10.3390/metabo9070123.
[9]   Keylor MH, Matsuura BS, Stephenson CR. Chemistry and biology of resveratrol-derived natural products. Chem Rev 2015, 115(17), 8976–9027.
[10] Přichystal J, Schug KA, Lemr K, Novák J, Havlíček V. Structural analysis of natural products. Anal Chem 2016, 88(21), 10338–10346.
[11] Rasul MG. Extraction, isolation and characterization of natural products from medicinal plants. Int J Basic Sci Appl Comput 2018, 2(6), 1–6.
[12] Zhang L. Comparison of extraction effect of active ingredients in traditional Chinese medicine compound preparation with two different method. Heilongjiang Xumu Shouyi 2013, 9, 132–133.
[13] Santos MCP, Gonçalves ÉCBA. Effect of different extracting solvents on antioxidant activity and phenolic compounds of a fruit and vegetable residue flour. Sci Agropecu 2016, 7, 7–14.
[14] Azwanida NN. A review on the extraction methods use in medicinal plants, principle, strength and limitation, Med Aromat Plants 2015, 4, 196. 10.4172/2167-0412.1000196.
[15] Wang L, Weller C. Recent advances in extraction of nutraceuticals from plants. Trends Food Sci Technol 2006, 17, 300–312.
[16] Vieitez I, Maceiras L, Jachmanián I, Alborés S. Antioxidant and antibacterial activity of different extracts from herbs obtained by maceration or supercritical technology. J Supercrit Fluids 2018, 133, 58–64.

[17] Phrompittayarat W, Putalun W, Tanaka H, Jetiyanon K, Wittaya-Areekul S, Ingkanina K. Comparison of various extraction methods of *Bacopa monnieri*. Naresuan Univ J 2007, 15(1), 29–34.

[18] Sasidharan S, Darah I, Jain K. *In vivo* and *in vitro* toxicity study of *Gracilaria changii*. Pharm Biol 2008, 46, 413–417.

[19] Cunha IBS, Sawaya ACHF, Caetano FM, Shimizu MT, Marcucci MC, Drezza FT, Povia GS, Carvalho PO. Factors that influence the yield and composition of Brazilian propolis extracts. J Braz Chem Soc 2004, 15(6), 964–970.

[20] Woisky RG, Salatino A. Analysis of propolis: Some parameters and procedures for chemical quality control. J Apic Res 1998, 37(2), 99–105.

[21] Zygmunt B, Namiesnik J. Preparation of samples of plant material for chromatographic analysis. J Chromatogr Sci 2003, 41(3), 109–116.

[22] Huie CW. A review of modern sample-preparation techniques for the extraction and analysis of medicinal plants. Anal Bioanal Chem 2002, 373, 23–30.

[23] Fu M, Zhang L, Han J, Li J. Optimization of the technology of ethanol extraction for Goupi patch by orthogonal design test. Zhongguo Yaoshi 2008, 11(1), 75–76.

[24] Gao X, Han J, Dai H, Xiang L. Study on optimizing the technological condition of ethanol percolating extraction for Goupi patch. Zhongguo Yaoshi 2009, 12(10), 1395–1397.

[25] Bandar H, Hijazi A, Rammal H, Hachem A, Saad Z, Badran B. Techniques for the Extraction of Bioactive Compounds from Lebanese *Urtica dioica*. Am J Phytomed Clin Ther 2013, 2321–2748.

[26] Rehman MU, Abdullah KF, Niaz K. Introduction to natural products analysis. In: Silva AS, Nabavi SF, Saeedi M, Nabavi SM, eds, Recent advances in natural products analysis. 1st, Amsterdam, The Netherlands, Elsevier, 2020, 3–15.

[27] Ciriminna R, Carnaroglio D, Delisi R, Arvati S, Tamburino A, Pagliaro M. Industrial feasibility of natural products extraction with microwave technology. ChemistrySelect 2016, 1(3), 549–555.

[28] Doktycz SJ, Suslick KS. Interparticle collisions driven by ultrasound. Science 1990, 247(4946), 1067–1069.

[29] Bucar F, Wube A, Schmid M. Natural product isolation -how to get from biological material to pure compounds. Nat Prod Rep 2013, 30(4), 525–545.

[30] Dai J, Mumper RJ. Plant phenolics: Extraction, analysis and their antioxidant and anticancer properties. Molecules 2010, 15(10), 7313–7352.

[31] Repajić M, Cegledi E, Kruk V, Pedisić S, Çınar F, Bursać Kovačević D, Žutić I, Dragović-Uzelac V. Accelerated solvent extraction as a green tool for the recovery of polyphenols and pigments from wild nettle leaves. Processes 2020, 8(7), 803. 10.3390/pr8070803.

[32] Mottaleb MA, Sarker SD. Accelerated solvent extraction for natural products isolation. Methods Mol Biol 2012, 864, 75–87.

[33] Sahena F, Zaidul IMZ, Jinap S, Karim AA, Abbas KA, Norulaini NAN, Omar AKM. Application of supercritical CO_2 in lipid extraction-A review. J Food Eng 2009, 95(2), 240–253.

[34] Rpff DS, Tap R-S, Duarte AC. Supercritical fluid extraction of bioactive compounds. Trends Analyt Chem 2016, 76, 40–51.

[35] Dilworth LL, Riley CK, Stennett DK. Plant constituents: Carbohydrates, oils, resins, balsams, and plant hormones. In: Mccreath SB, Delgoda R, eds, Pharmacognosy. 1st, London, UK, Academic Press, 2017, 61–80.

[36] Elyemni M, Louaste B, Nechad I, Elkamli T, Bouia A, Taleb M, Chaouch M, Eloutassi N. Extraction of essential oils of *Rosmarinus officinalis* L. by two different methods: Hydrodistillation and microwave assisted hydrodistillation, Sci World J 2019, 2019, 3659432. 10.1155/2019/3659432.

[37] Puri M, Sharma D, Barrow CJ. Enzyme-assisted extraction of bioactives from plants. Trends Biotechnol 2012, 30(1), 37–44.

[38] Liu T, Sui X, Li L, Zhang J, Liang X, Li W, Zhang H, Fu S. Application of ionic liquids based enzyme-assisted extraction of chlorogenic acid from *Eucommia ulmoides* leaves. Anal Chim Acta 2016, 903, 91–99.

[39] Strati IF, Gogou E, Oreopoulou V. Enzyme and high pressure assisted extraction of carotenoids from tomato waste. Food Bioprod Process 2015, 94, 668–674.

[40] Mgbeahuruike EE, Vuorela H, Yrjönen T, Holm Y. Optimization of thin-layer chromatography and high-performance liquid chromatographic method for *Piper guineense* extracts. Nat Prod Commun 2018, 13(1), 25–28.

[41] Kumar GS, Jayaveera KN, Kumar CKA, Sanjay UP, Swamy BMV, Kumar DVK. Antimicrobial effects of Indian medicinal plants against acne-inducing bacteria. Trop J Pharm Res 2007, 6(2), 717–723.

[42] Ám M, Ott PG, Yüce I, Darcsi A, Béni S, Morlock GE. Effect-directed analysis via hyphenated high-performance thin-layer chromatography for bioanalytical profiling of sunflower leaves. J Chromatogr A 2018, 1533, 213–220.

[43] Ansari S, Maaz M, Ahmad I, Hasan SK, Bhat SA, Naqui SK, Husain M. Quality control, HPTLC analysis, antioxidant and antimicrobial activity of hydroalcoholic extract of roots of *qust* (*Saussurea lappa*, C.B Clarke). Drug Metab Pers Ther 2020, 10.1515/dmdi-2020-0159.

[44] McAlpine JB, Friesen JB, Pauli GF. Separation of natural products by countercurrent chromatography. Methods Mol Biol 2012, 864, 221–254.

[45] Bojczuk M, Żyżelewicz D, Hodurek P. Centrifugal partition chromatography – A review of recent applications and some classic references. J Sep Sci 2017, 40(7), 1597–1609.

[46] Nahar L, Sarker SD. Droplet counter current chromatography (DCCC) in herbal analysis. Trends Phytochem Res 2020, 4(4), 201–202.

[47] Adams MA, Nakanishi K. Selected uses of HPLC for the separation of natural products. J Liq Chromatogr 1979, 2(8), 1097–1136.

[48] Swartz M. HPLC detectors: A brief review. J Liq Chromatogr Rel Technol 2010, 33, 1130–1150.

[49] Rahman A. Bioactive natural products (Part G), studies in natural products chemistry. Amsterdam, NL, Elsevier Science, 2002.

[50] Boligon AA, Athayde ML. Importance of HPLC in analysis of plants extracts. Austin Chromatogr 2014, 1(3), 2.

[51] Alvarez-Rivera G, Ballesteros-Vivas D, Parada-Alfonso F, Ibañez E, Cifuentes A. Recent applications of high resolution mass spectrometry for the characterization of plant natural products. Trends Anal Chem 2019, 112, 87–101.

[52] Malejko J, Nalewajko-Sieliwoniuk E, Szabuńko J, Nazaruk J. Ultra-high performance liquid chromatography with photodiode array and chemiluminescence detection for the determination of polyphenolic antioxidants in *Erigeron acris* L. Extracts Phytochem Anal 2016, 27(5), 277–283.

[53] Silva R, Lopes NP, Silva DB. Application of MALDI mass spectrometry in natural products analysis. Planta Med 2016, 82(8), 671–689.

[54] Bobzin SC, Yang S, Kasten TP. Application of liquid chromatography-nuclear magnetic resonance spectroscopy to the identification of natural products. J Chromatogr B Biomed Sci Appl 2000, 748(1), 259–267.

[55] Zhao W, Fernando LD, Kirui A, Deligey F, Solid-state WT. NMR of plant and fungal cell walls: A critical review, Solid State Nucl Magn Reson 2020, 107, 101660. 10.1016/j.ssnmr.2020.101660.

[56] Deborde C, Moing A, Roch L, Jacob D, Rolin D, Giraudeau P. Plant metabolism as studied by NMR spectroscopy. Prog Nucl Magn Reson Spectrosc 2017, 102–103, 61–97.

[57] Ma H, Bai Y, Li J, Chang YX. Screening bioactive compounds from natural product and its preparations using capillary electrophoresis. Electrophoresis 2018, 39(1), 260–274.

[58] Stashenko EE, Martínez JRGC-MS. Analysis of volatile plant secondary metabolites. In: Salih B, Çelikbıçak Ö, eds, Gas chromatography in plant science, wine technology, toxicology and some specific applications. Rijeka, Croatia, In Tech, 2012, 247–270.

[59] Al-Rubaye AF, Hameed IH, Kadhim MJ, Review: A. Uses of gas chromatography-mass spectrometry (GC-MS) technique for analysis of bioactive natural compounds of some plants. Int J Toxicol Pharmacol Res 2017, 9(1), 81–85.

[60] Poole CF. Conventional detectors for gas chromatography. In: Poole CF, ed., Handbooks in separation science, gas chromatography. 2nd, Amsterdam, The Netherlands, Elsevier, 2021, 343–369.

[61] Huck CW. Advances of infrared spectroscopy in natural product research. Phytochem Lett 2015, 11, 384–393.

[62] Hubert J, Nuzillard J-M, Renault J-H. Dereplication strategies in natural product research: How many tools and methodologies behind the same concept?. Phytochem Rev 2017, 16(1), 55–95.

[63] Lang G, Mayhudin NA, Mitova MI, Sun L, van der Sar S, Blunt JW, Cole ALJ, Ellis G, Laatsch H, Munro MHG. Evolving trends in the dereplication of natural product extracts: New methodology for rapid, small-scale investigation of natural product extracts. J Nat Prod 2008, 71, 1595–1599.

[64] Salem MA, Perez de Souza L, Serag A, Fernie AR, Farag MA, Ezzat SM, Alseekh S. (2020). Metabolomics in the context of plant natural products research: From sample preparation to metabolite analysis. Metabolites 2020, 10(1), 37. 10.3390/metabo10010037.

[65] Zani CL, Carroll AR. Database for rapid dereplication of known natural products using data from ms and fast nmr experiments. J Nat Prod 2017, 80, 1758–1766.

Index

https://doi.org/10.1515/9783110595949-012

www.ingramcontent.com/pod-product-compliance
Lightning Source LLC
Chambersburg PA
CBHW061411210326
41598CB00035B/6171